The Properties of Star-Forming Galaxies at z~2: Kinematics, Stellar Populations, and Metallicities

by

Dawn K. Erb

ISBN: 1-58112- 297-7

DISSERTATION.COM

Boca Raton, Florida
USA • 2005

The Properties of Star-Forming Galaxies at z~2:
Kinematics, Stellar Populations, and Metallicities

Dissertation.com
Boca Raton, Florida
USA • 2005

ISBN: 1-58112- 297-7

The Properties of Star-Forming Galaxies at $z \sim 2$: Kinematics, Stellar Populations, and Metallicities

Thesis by
Dawn K. Erb

In Partial Fulfillment of the Requirements
for the Degree of
Doctor of Philosophy

California Institute of Technology
Pasadena, California

2006

(Defended August 3, 2005)

Acknowledgements

More people than I can name here have helped to make the past five years happy and productive. My officemates at Caltech, Alison Farmer, Naveen Reddy, Jackie Kessler, and Alicia Soderberg, have always provided just the right amount of amusement and distraction. Patrick Shopbell, Anu Mahabal, and Cheryl Southard provided patient and knowledgeable technical support; and my especially heartfelt thanks to Patrick, Anu, Cheryl, and the makers of the DiskWarrior data recovery software, for saving me from having to repeat a significant amount of the work in this thesis. Thanks to P & L, and to Scott, Tatiana, Krista, Nancy, Alicia and my other teachers and friends at Yoga House, for always providing a welcoming, challenging and relaxing escape from the pressures of thesis-writing: namaste. Karin Menendez-Delmestre has been good company at home this last year. I have great appreciation for my collaborators in exploration of the $z \sim 2$ universe: Max Pettini, for being a great source of both encouragement and information, and for welcoming me on my visits to Cambridge; Kurt Adelberger, for always insightful comments and good though infrequent company; Naveen Reddy, for coming through all this with me from the beginning; and Alice Shapley, for her endless patience with my questions during the first years of this work, and for being a great observing partner on all those NIRSPEC runs. I hope there are more to come. And finally, I am especially grateful to have found one of the best advisors at Caltech in Chuck Steidel, who has always been encouraging, available, and patient with questions in spite of many other duties, and interested in any detail without ever losing sight of the big picture. He has always managed to strike just the right balance between showing concern with day to day operations and allowing independence. I am lucky to have collaborators who are such good people as well as good scientists. None of this would have happened without the support and encouragement of my parents, Dick and Judy Erb; the beginning of the belief that I could do this at all comes from them. And finally, to David Kaplan, for everything.

Abstract

We study the properties of star-forming galaxies at redshift $z \sim 2$, an era in which a substantial fraction of the stellar mass in the universe formed. Using 114 near-IR spectra of the Hα and [N II] emission lines and model spectral energy distributions fit to rest-frame UV through IR photometry, we examine the galaxies' star formation properties, dynamical masses and velocity dispersions, spatially resolved kinematics, outflow properties, and metallicities as a function of stellar mass and age. While the stellar masses of the galaxies in our sample vary by a factor of ~ 500, dynamical masses from Hα velocity dispersions and indirect estimates of gas masses imply that the variation of stellar mass is due as much to the evolution of the stellar population and the conversion of gas into stars as to intrinsic differences in the total masses of the galaxies. About 10% of the galaxies are apparently young starbursts with high gas fractions, caught just as they have begun to convert large amounts of gas into stars. Using the [N II]/Hα ratio of composite spectra to estimate the average oxygen abundance, we find a monotonic increase in metallicity with stellar mass. From the estimated gas fractions, we conclude that the observed mass-metallicity relation is primarily driven by the increase in metallicity as gas is converted to stars. The picture that emerges is of galaxies with a broad range in stellar population properties, from young galaxies with ages of a few tens of Myr, stellar masses $M_\star \sim 10^9$ M$_\odot$, and metallicities $Z \sim 1/3\,Z_\odot$, to massive objects with $M_\star \sim 10^{11}$ M$_\odot$, $Z \sim Z_\odot$, and ages as old as the universe allows. All, however, are rapidly star-forming, power galactic-scale outflows, and have masses in gas and stars of at least $\sim 10^{10}$ M$_\odot$, in keeping with their likely role as the progenitors of elliptical galaxies seen today.

Contents

List of Figures

List of Tables

Chapter 1

Introduction

Several lines of recent evidence indicate that the redshift range $1.5 \lesssim z \lesssim 2.5$ was one of the most active periods in the history of the universe. High levels of quasar activity and accretion onto massive black holes are indicated by the number density of luminous QSOs, which reaches a peak in this redshift range (e.g., Fan et al. 2001; Di Matteo et al. 2003), while the very bright but dusty submillimeter galaxies have recently been found to have a median redshift $z = 2.2$ (Chapman et al. 2005). Other studies indicate that much of the stellar mass in the universe today formed approximately during this redshift interval; Dickinson et al. (2003) find that 50–75% of the mass in today's galaxies had formed by $z \sim 1$, but only 3–14% had formed by $z \sim 2.7$. Similarly, Rudnick et al. (2003) find that the universe at $z \sim 3$ had $\sim 1/10$ the stellar mass density it has today, while half of the stellar mass in the universe had formed by $z \sim 1$–1.5. This era also sees the transition from the compact, irregular galaxies seen at $z \sim 3$ (Giavalisco et al. 1996) to the familiar Hubble sequence, which takes shape by $z \sim 1$.

Until recently, however, very little was known about the normal galaxies that populated the universe during this period. This span of cosmic time was known as the "redshift desert," because the strong emission lines usually used to identify galaxies shift out of the optical window past $z \sim 1.4$. However, such galaxies can in fact be selected efficiently by their rest-frame optical colors, through a method analogous to the selection of the Lyman break galaxies (LBGs) at $z \sim 3$. Their redshifts can be confirmed not through their optical emission lines, as has been traditional, but through their rest-frame UV spectra, which present a distinctive set of strong interstellar metal absorption lines. Advances in CCD sensitivity at short wavelengths have made such spectra relatively easy to obtain, as a typical $z \sim 2$ galaxy with $\mathcal{R} \sim 24$–25 can be well-detected with moderate ~ 1–2 hour exposure times on a blue-sensitive multi-object spectrograph attached to an 8–10 m telescope. In this way we have assembled a sample of ~ 1000 galaxies with redshifts $1.5 \lesssim z \lesssim 2.6$. This thesis is concerned with the properties of these galaxies. We describe the selection of the sample, and the methods used to study the galaxies, in more detail below.

1.1 The Selection of Galaxies at $z \sim 2$

The selection criteria we use for the $z \sim 2$ galaxies are adapted from those used to identifiy the Lyman break galaxies at $z \sim 3$ (Steidel et al. 2003), which have distinctive colors in the $U_nG\mathcal{R}$ filter set as the Lyman break at 912 Å passes through the U_n filter. There is no comparable break with which to identify galaxies at $z \sim 2$, but nevertheless such objects occupy a unique region in the $U_nG\mathcal{R}$ plane. The color criteria used to select these galaxies were developed by using simple models to predict the colors that galaxies with SEDs similar to the LBGs would have at $z \sim 2$, and then empirically verified and adjusted to improve their efficiency. Overviews of the $z \sim 2$ selection techniques and spectroscopic sample are given by Adelberger et al. (2004) and Steidel et al. (2004), respectively. We show the $U_nG\mathcal{R}$ color selection window for galaxies with $2 \lesssim z \lesssim 2.5$ in the left panel of Figure 1.1 (from Adelberger et al. 2004), with the predicted locations of galaxies at $z = 2$ and $z = 2.5$ shown by red and blue symbols. A histogram of the spectroscopic redshifts of galaxies selected by the $z \sim 2$ criteria is shown in the right panel. We use two sets of criteria aimed at galaxies with $1.5 \lesssim z \lesssim 2$ and $2 \lesssim z \lesssim 2.5$; most of our observing time has focused on candidates described by the higher redshift set. The surface density of objects satisfying either of the two criteria is ~ 9 arcmin^{-2}, and the fraction of low-redshift interlopers selected by the $2 \lesssim z \lesssim 2.5$ criteria is $\sim 9\%$, of which 3% are stars and 6% are star-forming galaxies at $z \sim 0.2$ (Steidel et al. 2004). Most interlopers are bright in the near-IR, with $K \lesssim 20$.

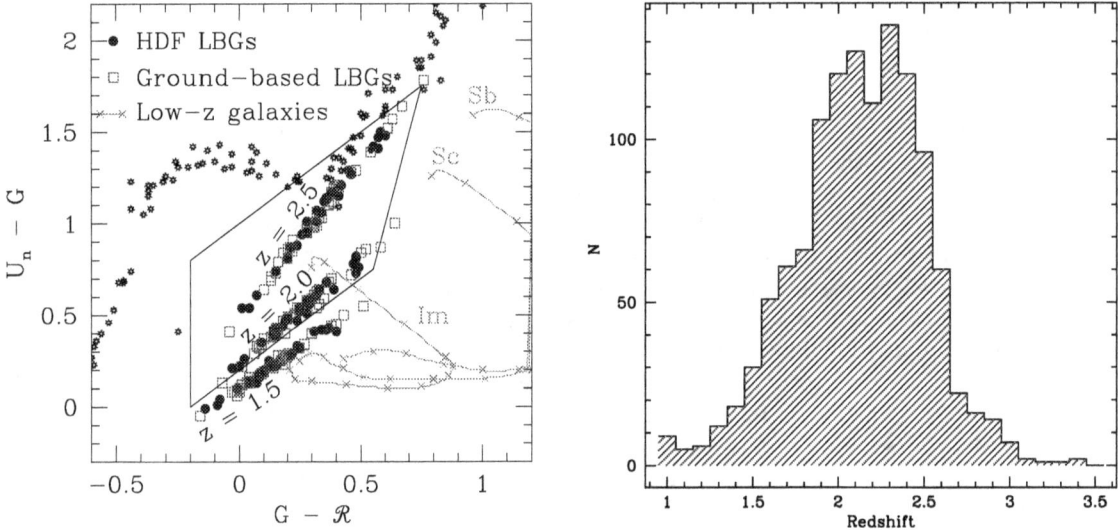

Figure 1.1 *Left:* The $U_n - G$ vs. $G - \mathcal{R}$ plane in which we select galaxies. The selection window for $2 \lesssim z \lesssim 2.5$ galaxies is shown by the trapezoid. It is primarily populated by star-forming galaxies in this redshift range, whose predicted locations (from de-redshifting the SEDs of the $z \sim 3$ LBGs) at $z = 2$ and $z = 2.5$ are shown by the red and blue symbols. A small fraction of interlopers, shown by stars (representing stars) and green tracks representing low-redshift galaxies, also enter the box. *Right:* A histogram of the spectroscopic redshifts of galaxies selected by the $z \sim 2$ critera. The distribution has $\langle z \rangle = 2.13 \pm 0.37$. 1198 galaxies are shown, 1031 of which have $1.5 < z < 2.6$.

No set of selection criteria will provide a complete sample, of course. It is not possible to use a single method to find all galaxies in a particular redshift range with reasonable efficiency; some balance must be struck between the fraction of interlopers allowed and the number of galaxies in the desired range rejected. Our $U_n G \mathcal{R}$ criteria miss all galaxies fainter than the $\mathcal{R} < 25.5$ cutoff; such objects may have very little current star formation or have nearly all their rest-frame UV light absorbed by dust. We also miss brighter highly reddened star-forming galaxies, which fall to the right of our selection window in Figure 1.1. Expanding the criteria to include these objects would substantially increase the interloper fraction as well.

Other criteria efficient at selecting galaxies at $z \sim 2$ have recently been developed. These are based on rest-frame optical colors, and are aimed at both star-forming (Daddi et al. 2004) and passive (Franx et al. 2003; Daddi et al. 2004) objects. Reddy et al. (2005) have recently quantified the overlap between the various samples, finding that galaxies selected by the $U_n G \mathcal{R}$ criteria account for $\sim 70\%$ of the star formation rate density at $z \sim 2$. The prime advantages of the rest-frame UV criteria are the modest amounts of observing time required (relative to the time needed to obtain deep K-band images for rest-frame optical selection), and, especially, the focus on galaxies that are bright enough to examine spectroscopically. It is primarily from such spectra that we gain insights into their properties.

1.2 The Importance of Rest-Frame Optical Spectra

The bulk of this thesis concerns the interpretation of a large sample of rest-frame optical spectra of the $z \sim 2$ galaxies, primarily focused on the Hα emission line. For $2 \lesssim z \lesssim 2.5$, H$\alpha$ falls in the K-band, [O III] and Hβ in the H-band, and [O II] in the J-band. The measurement of the full set of lines for a large sample of objects requires a prohibitive amount of observing time, however, because most near-IR spectrographs can observe only one band at a time, and efficient multi-object near-IR spectrgraphs are still being developed. We have therefore concentrated on Hα, assembling a sample of 114 spectra.

Because it is produced in H II regions photoionized by massive stars, the Hα luminosity is directly related to the number of massive stars and is therefore commonly used as a diagnostic of the star formation rate (SFR). It is among the most well-calibrated of such indicators (e.g., Kennicutt 1998a; Charlot & Longhetti 2001; Brinchmann et al. 2004), and the dependence of the conversion between Hα luminosity and SFR on metallicity and other factors is relatively well-understood. For this reason observations of Hα are particularly valuable for determining the star formation properties of high redshift galaxies; such data minimize the systematic uncertainties involved in the comparisons of local and distant samples. Hα is arguably more useful than, for example, the rest-frame UV continuum, which is much more easily obtained for large samples at $z \gtrsim 2$ (requiring only optical photometry), because UV observations of local galaxies are more difficult.

Hα is also an extremely useful tracer of galaxies' kinematics. It is used to determine rotation curves locally and up to $z \sim 1$, and provides a measure of a galaxy's dynamical mass through the Hα velocity dispersion. Not surprisingly, the determination of the kinematic properties of high redshift galaxies presents fundamental problems not encountered in the local universe; the lack of a priori knowledge of the galaxies' structure, and the lack of spatial resolution in the spectra, mean that kinematic data, whether spatially resolved or not, are difficult to interpret. Thus observations of Hα in local starburst galaxies are particularly useful for our present purposes, as these systems are probably the closest analogs to the $z \sim 2$ galaxies in our sample. Such data at least suggest the extent to which Hα may trace the potential wells of starbursts (e.g., Lehnert & Heckman 1996), and at least one recent study of Hα in local starbursts directly addresses the issues likely to be encountered at high redshift (Colina et al. 2005).

Because they allow a determination of the systemic redshifts of distant galaxies, spectroscopic observations of Hα (or the other nebular emission lines) in combination with the rest-frame UV allow measurements of the outflow speeds of the galactic-scale winds that are ubiquitous in starburst galaxies at both low and high redshifts. Offsets of several hundred km s^{-1} are seen between the redshifts of the nebular lines and the interstellar absorption lines, which are typically blueshifted, and between the nebular lines and Lyα, which is redshifted (Pettini et al. 2001). These offsets are interpreted through the standard model in which we see interstellar absorption from approaching outflowing material, while Lyα is redshifted through resonant scattering off the receding shell on the far side of the galaxy. The outflow velocities of winds in local starbursts have been seen to vary with the masses and star formation rates of their host galaxies (Martin 2005; Rupke et al. 2005), and the determination of the outflow velocities in a large sample of $z \sim 2$ galaxies in combination with the SFRs and dynamical masses from Hα allow us to test for the existence of such correlations at high redshift.

Finally, the ratio of [N II]$\lambda6583$ to Hα correlates with the abunance of oxygen and is therefore useful as an indicator of metallicity (Pettini & Pagel 2004). Because the [N II] line is quite weak in individual spectra, composite spectra that increase the S/N are the only way to determine the abundances of all but the most metal-rich galaxies in our sample (for which [N II] can be well-detected in individual spectra;

Shapley et al. 2004). A large sample, which can be subdivided according to properties that might be expected to vary with metallicity, therefore enables useful studies of abundance trends at high redshift. We carry out such a study in Chapter 4, dividing the sample of Hα spectra into bins by both stellar mass and luminosity.

1.3   Stellar Population Parameters from Model Spectral Energy Distributions

The other major component of this thesis involves the estimation of the stellar masses, ages, star formation rates, and reddening of the $z \sim 2$ galaxies via the comparison of multi-wavelength photometry and model spectral energy distributions. Such techniques are well-established in the study of high redshift galaxies (Papovich et al. 2001; Shapley et al. 2001, 2005b). Photometry extending at least to the rest-frame optical (the observed K-band) is required, and the addition of rest-frame IR data, which are more senstive to older stars and somewhat less influenced by current star formation, improves the uncertainties considerably (Shapley et al. 2005b). Stellar masses are the most well-constrained of the fitted parameters, while degeneracies between age and reddening, which have a similar effect on the UV slope, limit the accuracy to which the other properties can be determined. For this reason the most useful results come from large samples; even when the properties of individual objects are uncertain, statistical trends can provide useful results. We focus here on relating the properties determined through SED modeling to those derived from the Hα spectra, looking particularly at variations of star formation rate, dynamical mass, and metallicity with stellar mass.

The outline of the thesis is as follows. Chapter 2 describes our pilot observations of Hα at $z \sim 2$, which were conducted in May 2002. We present results for 16 galaxies and focus on their kinematics and star formation rates. This chapter has been published as Erb et al. (2003). The work in Chapter 3 (published as Erb et al. 2004) was inspired by the surprisingly large fraction of galaxies in our pilot study that showed spatially resolved and tilted emission lines suggestive of rotation; it describes a study of the kinematics of galaxies selected for their elongated morphologies. In Chapter 4 we discuss the full sample of 114 spectra, and study the variations of the properties determined from Hα (including Hα line width and dynamical mass, amplitude of velocity shear, outflow velocities, Hα equivalent width, and star formation rate) with stellar mass, age and rest-frame optical luminosity. We also introduce estimates of the galaxies' gas masses, derived from the empirical local correlation between star formation density and gas density. In Chapter 5 we construct composite near-IR spectra as a function of stellar mass and find a relation between stellar mass and metallicity, which we examine in the context of the simple, "closed box" model of chemical evolution using the gas masses mentioned above. A brief epilogue describes two ways that new telescope instrumentation may enable the resolution of questions that remain from this work.

Chapter 2

Hα Spectroscopy of Galaxies at $z > 2$: Kinematics and Star Formation[*][†]

Dawn K. Erb,[a] Alice E. Shapley,[a] Charles C. Steidel,[a‡] Max Pettini,[b]
Kurt L. Adelberger,[c§] Matthew P. Hunt,[a] Alan F. M. Moorwood,[d] & Jean-Gabriel Cuby[e]

[a]California Institute of Technology, Department of Astronomy, MS 105-24, Pasadena, CA 91125

[b]Institute of Astronomy, Madingley Road, Cambridge CB3 0HA, UK

[c]Harvard-Smithsonian Center for Astrophysics, 60 Garden Street, Cambridge, MA 02138

California Institute of Technology, Department of Astronomy, MS 105-24, Pasadena, CA 91125

[d]European Southern Observatory, Karl-Schwarzschild-Str. 2, D-85748 Garching, Germany

[e]European Southern Observatory, Alonso de Cordova 3107, Santiago, Chile

Abstract

We present near-infrared spectroscopy of Hα emission lines in a sample of 16 star-forming galaxies at redshifts $2.0 < z < 2.6$. Our targets are drawn from a large sample of galaxies photometrically selected and spectroscopically confirmed to lie in this redshift range. We have obtained this large sample with an extension of the broadband $U_nG\mathcal{R}$ color criteria used to identify Lyman break galaxies at $z \sim 3$. The primary selection criterion for IR spectroscopic observation was proximity to a QSO sightline; we therefore expect the galaxies presented here to be representative of the sample as a whole. Six of the galaxies exhibit spatially extended, tilted Hα emission lines; rotation curves for these objects reach mean velocities of ~ 150 km s^{-1} at radii of ~ 6 kpc, without corrections for inclination or any other observational effect. The velocities and radii give a mean dynamical mass of $\langle M \rangle \geq 4 \times 10^{10} M_\odot$. We have obtained archival *HST* images for two of these galaxies; they are morphologically irregular. One-dimensional velocity dispersions for the 16 galaxies range from ~ 50 to ~ 260 km s^{-1}, and in cases where we have both virial masses

[*]Based, in part, on data obtained at the W.M. Keck Observatory, which is operated as a scientific partnership among the California Institute of Technology, the University of California, and NASA, and was made possible by the generous financial support of the W.M. Keck Foundation. Also based in part on observations collected at the European Southern Observatory, Paranal, Chile (ESO Programme 66.A-0206).

[†]A version of this chapter was published in *The Astrophysical Journal*, vol. 591, 101–118.

[‡]Packard Fellow

[§]Harvard Society Junior Fellow

implied by the velocity dispersions and dynamical masses derived from the spatially extended emission lines, they are in rough agreement. We compare our kinematic results to similar measurements made at $z \sim 3$, and find that both the observed rotational velocities and velocity dispersions tend to be larger at $z \sim 2$ than at $z \sim 3$. We also calculate star formation rates (SFRs) from the Hα luminosities, and compare them with SFRs calculated from the UV continuum luminosity. We find a mean $SFR_{H\alpha}$ of 16 $M_\odot$ yr^{-1} and an average $SFR_{H\alpha}/SFR_{UV}$ ratio of 2.4, without correcting for extinction. We see moderate evidence for an inverse correlation between the UV continuum luminosity and the ratio $SFR_{H\alpha}/SFR_{UV}$, such as might be observed if the UV-faint galaxies suffered greater extinction. We discuss the effects of dust and star formation history on the SFRs, and conclude that extinction is the most likely explanation for the discrepancy between the two SFRs.

2.1   Introduction

Our knowledge of star-forming galaxies at high redshift has increased enormously in the past ten years, particularly at $z \sim 3$; large samples of galaxies at these redshifts are now known (Steidel et al. 1999, 2003), and they have been studied in both the rest-frame UV (Pettini et al. 2000; Shapley et al. 2003) and optical (Shapley et al. 2001; Papovich et al. 2001; Pettini et al. 2001), as well as at submillimeter (Chapman et al. 2000; Adelberger & Steidel 2000) and X-ray (Nandra et al. 2002) wavelengths to some extent. Much less is known about galaxies at $z \sim 2$. Because these objects lack strong spectroscopic features in the optical window, they have traditionally been difficult to identify. This is unfortunate, as $z \sim 2$ is likely the epoch in which a large fraction of the stars in the present day universe formed (Madau, Pozzetti, & Dickinson 1998; Blain et al. 1999), in which bright QSO activity reached its peak (Schmidt, Schneider, & Gunn 1995; Pei 1995; Fan et al. 2001), and in which rapidly star-forming galaxies of compact and disordered morphologies became the normal Hubble sequence galaxies of the $z < 1$ universe (Dickinson 2000).

The situation is improving, however. With the advent of sensitive IR detectors observations of rest-frame optical features are now feasible, and have been carried out successfully. Teplitz, Malkan, & McLean (1998) reported 11 Hα emitters discovered in a narrow-band IR imaging survey; Yan et al. (1999) and Hopkins, Connolly, & Szalay (2000) used slitless spectroscopy with the Near Infrared Camera and Multi-Object Spectrograph (NICMOS) on the *Hubble Space Telescope (HST)* to study the Hα luminosity function and star formation rate in galaxies at $z \leq 1.9$. Objects at $z \sim 2$ are in fact ideally suited for ground-based IR spectroscopy, since Hα falls in the K-band, [O III] and Hβ in the H-band, and [O II] in the J-band. This coincidence has been exploited with recent observations employing near-IR spectrographs on 8–10 m telescopes; most of these have focused on Hα emission (Kobulnicky & Koo 2000; Lemoine-Busserolle et al. 2003). Among these spectra is a rotation curve of a galaxy at $z \sim 2$ that reaches a velocity of $\gtrsim 200$ km s^{-1} (Lemoine-Busserolle et al. 2003), suggesting that near-IR spectroscopy may be able to provide the most detailed kinematic information yet available on galaxies at high redshift. It is also clear from most of the above results that star formation rates measured from Hα are consistently higher than those measured from the UV continuum luminosity; this is in accordance with observations at $z \sim 1$ (Glazebrook et al. 1999; Tresse et al. 2002) and at lower redshifts (e.g., Bell & Kennicutt 2001; also see Sullivan et al. (2000) and Buat et al. (2002) for comparisons of Hα and UV SFRs). The difference is generally accounted for by the differing sensitivities of the Hα and UV continuum star formation rate diagnostics to the presence of dust and to star formation history.

In this paper we present Hα spectroscopy in the K-band of 16 UV-selected galaxies in the redshift range $2.0 < z < 2.6$. In §2.2 we describe our target selection process, observations, and data reductions. In §2.3 we comment individually on any noteworthy features of the galaxies. Section 2.4 addresses the kinematics of the galaxies: we discuss the rotation curves in §2.4.1 and the one-dimensional velocity dispersions in §2.4.2. In §2.5 we calculate star formation rates from Hα and rest-frame UV emission and compare them, and we discuss our conclusions in §2.6. We use a cosmology with $H_0 = 70$ km s^{-1} Mpc^{-1}, $\Omega_m = 0.3$, and $\Omega_\Lambda = 0.7$ throughout. In this cosmology, the universe at $z = 2.3$ is 2.8 Gyr old, or 21% of its present age, and a proper distance of 8.2 kpc subtends an angular distance of $1''$.

2.2 Target Selection and Observations

The objects discussed herein are drawn from a large sample of galaxies photometrically selected and spectroscopically confirmed to be in the redshift range $2.0 \leq z \leq 2.6$. We summarize the selection technique here; a more complete discussion will be given in a forthcoming paper. We have extended the broadband color criteria used to select galaxies at $z \sim 3$ (Steidel & Hamilton 1993; Steidel, Pettini, & Hamilton 1995; Steidel et al. 1996) to other regions of the $(U_n - G)$ vs. $(G - \mathcal{R})$ plane, identifying candidates according to the following conditions:

$$
\begin{aligned}
G - \mathcal{R} &\geq -0.1 \\
U_n - G &\geq G - \mathcal{R} + 0.2 \\
G - \mathcal{R} &\leq 0.3(U_n - G) + 0.2 \\
U_n - G &< G - \mathcal{R} + 1.0.
\end{aligned}
\tag{2.1}
$$

We refer to these objects as "BX" (e.g., Q1700-BX691); 92% of the objects satisfying these criteria are galaxies in the redshift range $1.6 \leq z \leq 2.8$, with 72% in the range $2.0 \leq z \leq 2.6$. These criteria were developed by calculating the colors that typical $z \sim 3$ Lyman break galaxies (LBGs) would have if they were placed at $z \sim 2$; they are therefore designed to select objects with similar intrinsic spectral energy distributions (SEDs) at both redshifts (Adelberger 2002). Our sample also contains four "MD" objects (e.g., Q1623-MD107); these objects are detected in the U_n-band and meet the criteria

$$
\begin{aligned}
G - \mathcal{R} &< 1.2 \\
U_n - G &\leq G - \mathcal{R} + 1.5 \\
U_n - G &\geq G - \mathcal{R} + 1.0.
\end{aligned}
\tag{2.2}
$$

They have the redshift distribution $\langle z \rangle = 2.79 \pm 0.27$ (Steidel et al. 2003), so that the low redshift end of the distribution encompasses objects with $z \leq 2.6$. Both the BX and MD candidates are restricted to $\mathcal{R} \leq 25.5$ (roughly equivalent to $\mathcal{R} \lesssim 26$ at $z \sim 3$). The two remaining objects in our sample, Q0201-B13 and CDFb-BN88, satisfy the BX criteria but have different names because they predated the systematic use of the $z \sim 2$ selection technique. Once candidates are photometrically identified, we confirm their redshifts with rest-frame UV spectra obtained with the Low Resolution Imaging Spectrometer (LRIS; Oke et al. 1995) on the Keck I telescope. The redshifts from the UV interstellar absorption lines and Lyα when present are listed in Table 2.1, and spectra for two of the objects are shown as examples in Figure 2.1. The rest-frame UV observations will be described in detail elsewhere.

The galaxies targeted for IR spectroscopy were selected as part of an ongoing project examining the interplay between galaxies and the intergalactic medium (IGM) in which we combine spectroscopy of faint star-forming galaxies with QSO absorption line observations of the IGM in the same volume (Adelberger et al. 2003). A detailed comparison of the galaxies and the IGM requires accurate measurements of the galaxy redshifts, and ultimately an understanding of the star formation rates, masses, and ages of galaxies near the QSO lines of sight; therefore the primary selection criterion (beyond the color criteria described above) for the present sample was proximity to a QSO sightline. This naturally results in a sample with a wide range of UV properties (as distinguished, for example, from the galaxies in the $z \sim 3$ sample of Pettini et al. (2001), which were selected to be particularly UV-bright).

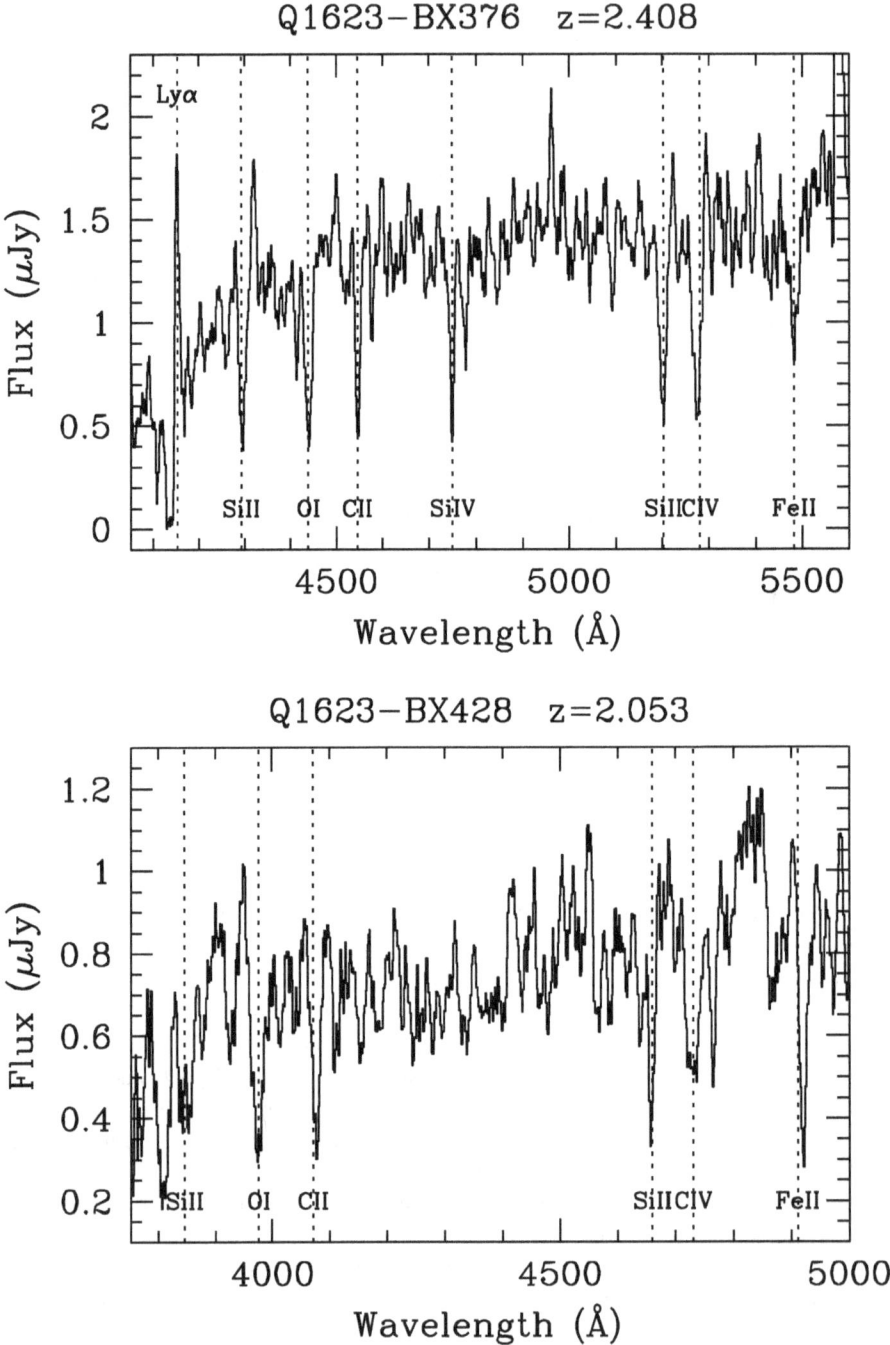

Figure 2.1 Sample rest-frame UV spectra for two of the galaxies. We show Q1623-BX376 at $z_{abs} = 2.408$ (top), and Q1623-BX428 at $z_{abs} = 2.053$ (bottom). The rest wavelengths of the lines labeled are Lyα λ1215 Å, SiII λ1260 Å, OI λ1302 Å, CII λ1334, SiIV λ1394, SiII λ1526, CIV λ1549, and FeII λ1608.

Twelve of our 16 galaxies are within $60''$ of QSOs in fields at 1700+64 and 1623+27, and have redshifts slightly lower than those of the QSOs themselves; these were observed with the Near Infrared Imaging Spectrograph (NIRSPEC; McLean et al. 1998) on the Keck II telescope in May 2002. We observed an additional galaxy in the Groth-Westphal field on the same run. The other three objects in the sample (SSA22a-MD41, Q0201-B13, and CDFb-BN88) were observed with the Infrared Spectrometer and Array Camera (ISAAC; Moorwood et al. 1998) on the Very Large Telescope 1 (VLT 1) in October 2000, and were among the small number of $z \sim$ 2–2.5 galaxies in the $z \sim 3$ LBG survey fields at the time. They were also selected because of their UV brightness, and because of the favorable wavelength of Hα relative to night sky emission lines and the possibility of measuring rotation.

2.2.1 Data Acquisition

Most of our targets were observed on May 19 and 20, 2002 (UT) with the NIRSPEC spectrograph on the Keck II telescope. NIRSPEC is described in detail by McLean et al. (1998); it uses a 1024 × 1024 pixel (ALADDIN2) InSb detector with 27 μm pixels. In the medium-dispersion mode used for these observations, each detector pixel corresponds to $0''\!.143$ in the spatial direction, and the dispersion in the spectral direction is 4.2 Å per pixel. We used a $0''\!.76 \times 42''$ entrance slit, which gives a resolving power of $R \simeq 1400$ corresponding to a spectral resolution of ~ 15 Å FWHM in the observed frame K-band, as measured from the widths of sky lines. In almost all cases we were able to place two galaxies on the slit at the same time by setting the appropriate position angle. Because the galaxies are too faint to be acquired directly on the spectrograph slit, we placed them on the slit by offsetting from a nearby bright star or from the QSO with a sightline near the galaxy. Individual exposures were 900 s, and we typically took four exposures of each object for a total of 1 hour of integration. Between each exposure we reacquired the offset star, moved it along the slit by approximately $5''$, and offset once again to the target object. The detector was read out in multiple-read mode, with 16 reads at the start and end of each integration; the results were then averaged to reduce noise. The choice of filter and wavelength range was governed by the expected position of the Hα line based on each galaxy's optical redshift; we used the NIRSPEC6 and NIRSPEC7 filters, which span the wavelength ranges 1.56–2.32 and 1.84–2.63 μm, respectively. The spectral dispersion allows a range of approximately 0.4 μm to be placed on the detector at one time. Conditions were photometric on both nights, with approximately $0''\!.5$ FWHM seeing in K-band.

SSA22a-MD41, Q0201-B13, and CDFb-BN88 were observed on October 20–22, 2000 (UT) with the ISAAC spectrograph on the VLT1. The short-wavelength channel of ISAAC (Moorwood et al. 1998) uses a 1024 × 1024 pixel Rockwell HgCdTe array with 18.5 μm pixels. The pixel scale along the $1'' \times 120''$ slit is similar to that of NIRSPEC, $0''\!.146$ pixel^{-1}, but the spectral resolution is 2.5 times higher, with $R \simeq 3500$ and sky line widths of ~ 6 Å FWHM. We observed in the K-band, again targeting the expected position of Hα from rest-frame UV redshifts. The position angles were chosen to align with the major axes of the galaxies if any extended structure was apparent in the optical images; this was the case with SSA22a-MD41 and with CDFb-BN88 to a lesser extent. We also placed a bright star on the slit along with each galaxy to facilitate the determination of offsets between images. We performed an ABBA series of four 720 second exposures, with $10''$ offsets between the A and B positions. The object was then reacquired at a different position along the slit and the procedure was repeated, typically for a total of ~ 3 hours of integration. Conditions were not photometric, and the seeing varied between $0''\!.5$ and $0''\!.6$ FWHM. The targets and observations are summarized in Table 2.1.

2.2.2 Data Reduction

The fully reduced spectra are shown in Figure 2.2. The two-dimensional images were reduced with IRAF; preliminary steps included flagging and masking any pixels that exhibited aberrant behavior in the dark and flat-field images, flat-fielding the data using the spectrum of a quartz halogen lamp, and cutting out and rotating the image of the slit. Spatial distortion was corrected by stepping a bright star along the slit at 5″ intervals for the NIRSPEC data and 10″ intervals for that from ISAAC, combining the resulting images, and determining the star trace as a function of slit position. We then applied a wavelength solution to the rectified images by identifying the OH sky lines with reference to a list of vacuum wavelengths from the Kitt Peak National Observatory Fourier Transform Spectrograph,[1] resulting in 2-D images rectified both spatially and spectrally.

For the NIRSPEC objects, we took four 900 s exposures of each galaxy or galaxy pair, moving the object(s) along the slit for each integration. In order to subtract the sky background, we constructed a sky frame from the temporally adjacent images; after scaling and smoothing in the spatial direction, this sky frame was subtracted from the science image. Sky subtraction was done slightly differently for the ISAAC observations, which were taken using ABBA offsets: A sky frame made from the sum of the A images was subtracted from the B images, and vice versa. Further background subtraction was done for both the NIRSPEC and ISAAC observations by fitting a polynomial in the spatial direction at each wavelength bin, avoiding the positions of any bright objects on the slit; this removed some of the residuals of the sky lines. Finally, we produced a fully reduced, two-dimensional spectrogram of each galaxy by registering and averaging the individual frames, excluding bad pixels identified from combined dark and flat-field images. This step also produced a two-dimensional frame of the statistical 1 σ error appropriate to each pixel. The last step was to extract one-dimensional spectra of each galaxy; this was done by summing the pixels containing a signal along the slit. The same aperture was then used to extract a variance spectrum from the square of the error image described above; the square root of this is a 1 σ error spectrum that was used to determine the uncertainties in the line fluxes and widths.

2.2.3 Flux Calibration

In order to put the one-dimensional spectra onto an absolute flux scale, we observed A0 and A2 stars from the list of UKIRT photometric standards.[2] These typically have $K \simeq 7$ mag, and were observed at similar airmass and with the same instrumental configuration as the galaxies themselves. Flux calibration was done by scaling the spectral energy distribution of Vega (Colina, Bohlin, & Castelli 1996) according to the magnitude of the standard used, and dividing the spectrum of the standard star by this scaled Vega spectrum. This gives a sensitivity function in counts per unit flux density, by which we divided our one-dimensional galaxy spectra. Because the spectra of A stars are relatively smooth at the wavelengths of interest, they provide a measurement of the atmospheric absorption, and dividing our galaxy spectra by the sensitivity function therefore corrects for atmospheric absorption.

The uncertainties in the flux calibration process are both substantial and difficult to quantify; however, we have attempted to estimate them in several ways. As described above, we extracted 1 σ error spectra for each of the galaxies; these primarily reflect the noise of the sky background. By integrating the flux in the variance (σ^2) spectrum at the position of Hα and taking the square root of the result, we can measure the random error associated with the observation; this is $\leq$10%. More difficult to measure are systematic errors: The largest sources of uncertainty are the flux lost due to imperfect centering of the objects on the slit, seeing and seeing variations, and the possibility of the objects being larger than the slit itself. We can get a sense of the importance of these effects by comparing the fluxes received in each of the individual exposures that were co-added to produce our final spectra. We find that flux levels between exposures vary by about 30% (1 σ); this includes random as well as systematic error. The uncertainty in the mean flux of our three or four exposures is then 15–20%. This accounts for variations in object centering and seeing, but not for flux consistently lost due to the width of the slit. As the galaxies observed are small

---

[1]Available at `http://www2.keck.hawaii.edu:3636/inst/nirspec/data/oh.lst`

[2]Available at `http://www.jach.hawaii.edu/JACpublic/UKIRT/astronomy/calib/`

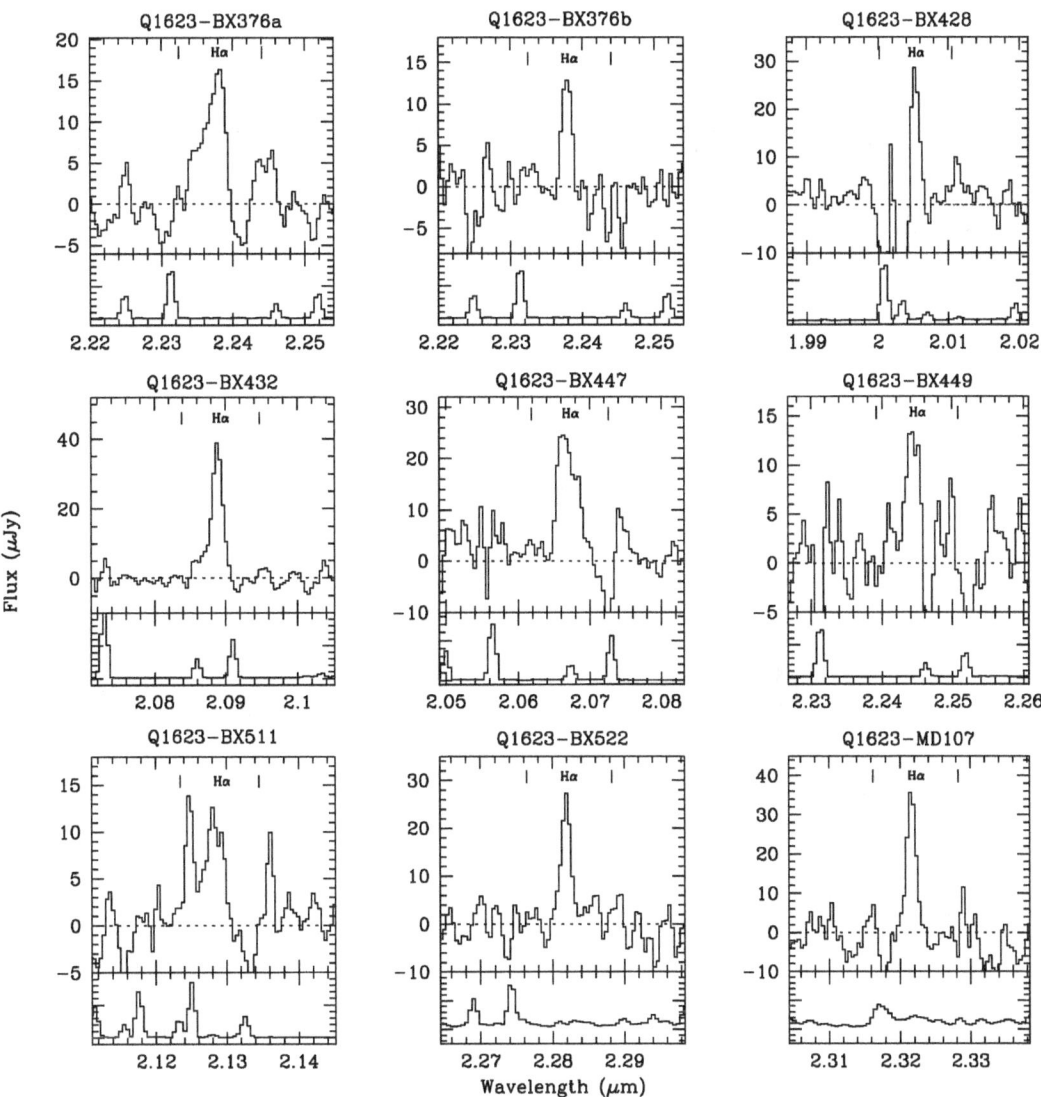

Figure 2.2 Fully reduced one-dimensional spectra for all of the galaxies in our sample. The Hα emission line is marked on each spectrum, and the vertical lines to either side mark the positions at which [N II] emission would appear. Plotted below each galaxy spectrum is a sky spectrum, in arbitrary flux units. The spectra have been smoothed with a two pixel boxcar filter. We plot a larger wavelength range for Q1700-BX691, the only object in which we see [N II] and [S II] emission. The last three objects, SSA22a-MD41, Q0201-B13, and CDFb-BN88, were observed with ISAAC on the VLT, and their spectra have been smoothed to approximate the resolution of NIRSPEC. We discuss the objects individually in § 2.3.

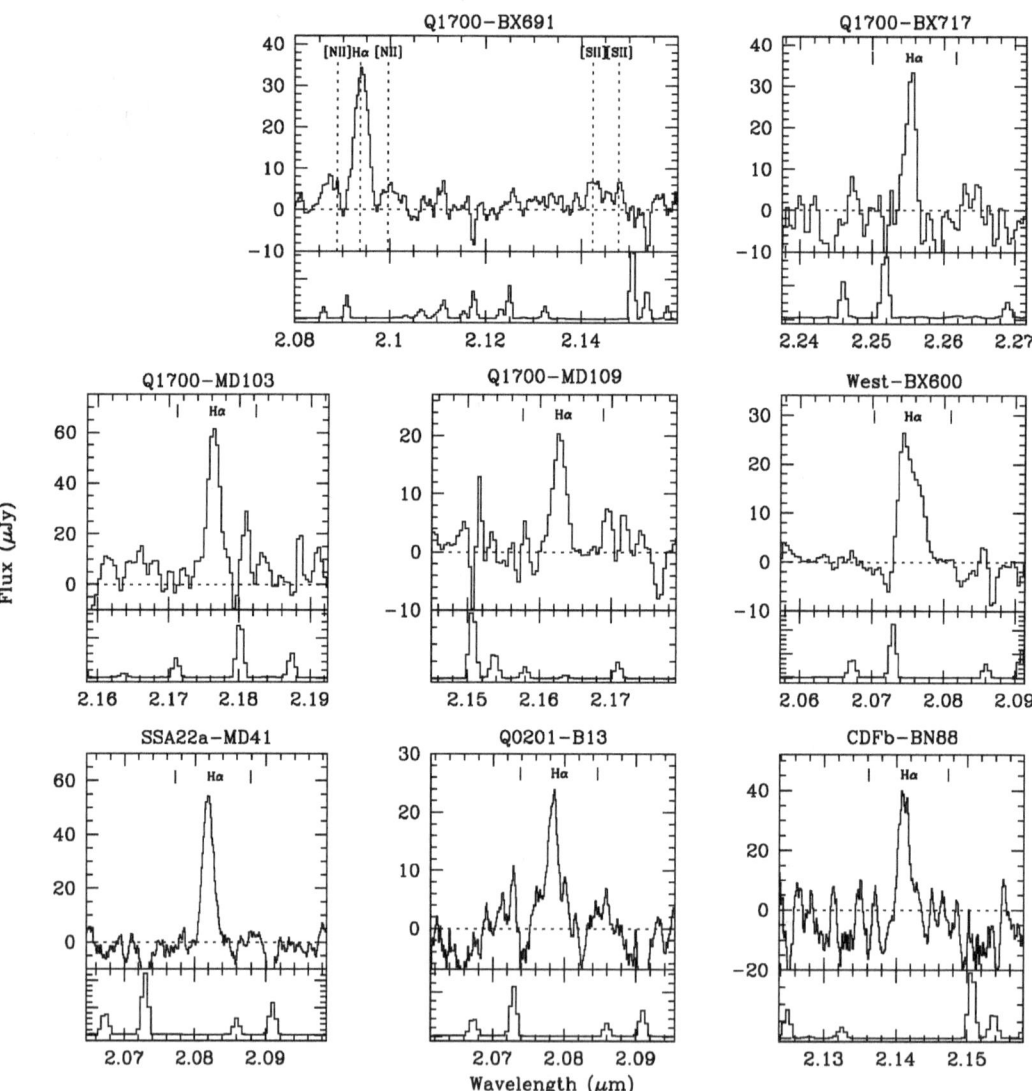

Figure 2.2 Continued

$(r_{1/2} \sim 0\rlap{.}''2\text{--}0\rlap{.}''3$ in an *HST* WFPC2 pointing that includes several of them), we assume that in most cases the flux loss is not significant; however, a few of the galaxies are particularly irregular and extended, and in these cases the flux loss may be significant. We can perform a further check by calibrating the same object with several different standard stars; in doing so we find variations in flux of 15% at maximum, and usually much less (again, 1 σ). Because we have K'-band photometry for one of the galaxies in our sample (Q1700-BX691, one of the few in which we detect a continuum signal), we can compare the photometric flux with the continuum flux; we find that our spectrum underestimates the photometric flux by a factor of 1.3, or about 25%. Because the continuum is so faint, this measurement is subject to large errors, and is more a test of our sky subtraction than of our spectrophotometry. We have also extracted one-dimensional spectra of the standard stars with a variety of aperture widths in order to determine whether an aperture correction might be necessary; we find that less than 5% of the flux is lost with the apertures used to extract the galaxy spectra. As this is much smaller than other sources of error, no aperture correction was applied. Based on all of these tests, we take our measured fluxes as uncertain by about 25%. This uncertainty propagates directly into the derived luminosities and star formation rates, and will be adopted in the analyses that follow.

2.3 Comments on Individual Objects

While our selection process naturally leads to a wide range of UV properties, with $\mathcal{R}$ ranging between 23.1 and 25.5 mag (i.e., a factor of 9 in luminosity), we see less variation in the Hα fluxes, which vary only by a factor of 4. The UV and Hα properties are not necessarily correlated, however; some of the strongest Hα luminosity comes from the faintest UV objects. Because we have rest-frame UV spectra of the galaxies, we are confident that none of them are AGN; we see no high-ionization emission lines, and few even show Lyα emission. The lack of strong [N II] emission also indicates that the galaxies are not AGN. Veilleux & Osterbrock (1987) find that AGN have log [N II]λ6583/H$\alpha \gtrsim -0.2$, while we detect [N II] emission in only one case, and that weakly. We hesitate to infer anything about the metallicity of the galaxies based on the absence of these lines, however, given the limited S/N of our data. The galaxies are faint, with no spectroscopically detected continuum in most cases, and we have not yet obtained rest-frame optical magnitudes; therefore we are unable to calculate Hα equivalent widths. We see a larger variation in the velocity dispersion σ than in previous samples of comparable size at high redshift (Pettini et al. 2001), but the most notable feature of our sample is the six galaxies that show evidence of ordered rotation, as we discuss in § 2.4.1. We comment on each object below.

Q1623-BX376: This is one of the brightest rest-frame UV objects in our sample, and the only one in which the star formation rate calculated from the UV emission is unambiguously higher than that from the Hα emission (see § 2.5). In ground-based imaging it appears extended, with a fainter component extending $\sim 2\rlap{.}''5$ to the west. The association between the two components is less clear with higher-resolution imaging (see Figure 2.3); however, the Hα emission also consists of two lines at the same redshift, separated by $2\rlap{.}''5$. We have extracted spectra for both components, as shown in Figure 2.2; the primary component is labeled Q1623-BX376a, and the fainter Q1623-BX376b. Because our optical photometry treated both components as a single extended object, we sum the fluxes from both lines in order to calculate the Hα star formation rate in § 2.5.

Q1623-BX428: Unfortunately this galaxy lies at a redshift such that Hα falls very close to a strong sky line, to which we have lost significant flux. This can be seen clearly in Figure 2.2, where the sky line falls just to the left of Hα. Because of the loss of flux we are able to place a lower limit on the Hα star formation rate, but the sky subtraction has affected the line profile such that the velocity dispersion cannot be determined.

Q1623-BX447: This is one of the six galaxies for which we derived rotation curves from tilted Hα

emission lines; it is also one of the few for which we have *HST* imaging, which shows it to be morphologically complicated (see Figure 2.3). We also see from the *HST* image that our slit was offset from the most extended axis of the galaxy by ~ 60 degrees.

Q1623-BX511: Of the six galaxies for which we were able to derive rotation curves, this has the smallest Hα flux and hence the smallest spread in velocity and the largest uncertainties. The Hα emission falls between two bright sky lines, as can be seen in Figure 2.2. At $\mathcal{R}= 25.37$ it is among the faintest UV objects in our sample as well.

Q1700-BX691: This is the only galaxy in which we clearly detect [N II]$\lambda\lambda6549$, 6583 and [S II]$\lambda\lambda6717$, 6734 emission lines as well as Hα. All of the lines are tilted in the two-dimensional spectra, providing strong evidence for rotation. The Hα rotation curve reaches a velocity of ~ 240 km s^{-1} at ~ 9 kpc, with no sign of flattening; this is clearly a massive system. The fact that we see [N II] and [S II] lines suggests a relatively high metallicity; however, we defer a calculation until we are able to obtain measurements of [O III] in the *H*-band. Interestingly this is among the faintest UV objects in our sample, with $\mathcal{R}=25.33$. A K'-band image of this object (Teplitz et al. 1998; private communication) shows it to be extremely red, with $\mathcal{R}-K' = 5.10$. The K'-band image also shows that our slit was fortuitously aligned with the major axis.

Q1700-MD103: This galaxy has the strongest Hα emission in our sample, and hence the largest Hα-derived star formation rate, 27 M$_\odot$ yr^{-1}. It is also one of the six objects in which we detect rotation.

Westphal-BX600: One of the six objects in which we detect rotation, this galaxy is second only to Q1700-BX691 in rotational velocity and implied mass. We detected Hα emission serendipitously, while observing the nearby $z \sim 3$ galaxy Westphal-MD115. This object had been previously classified as a $z \sim 2$ galaxy candidate based on its rest-frame UV colors, but it has not yet been observed with LRIS. Although we have no optical redshift, we believe the line detected here to be Hα because its UV colors are entirely consistent with a redshift of $z = 2.16$; the contamination fraction in the optical color selection process is less than 10%, with most of the interlopers being galaxies at low redshift ($z = 0.05$–0.15). We do not know of any strong emission lines that would fall in our spectral window for a galaxy in this redshift range; for a redshift of $z = 0.008$, He I (2.058 μm; Lançon et al. 2001) would fall at the wavelength of the observed line, but then we would also expect to see stronger Brγ (2.166 μm) emission at 2.18 μm, which we do not.

SSA22a-MD41: This is one of the three galaxies that were observed with the ISAAC spectrograph on the VLT. Conditions were not photometric during the ISAAC run, so we are only able to place a lower limit on the Hα star formation rate. We detect rotation in the Hα emission, with a large spatial extent of nearly ± 10 kpc.

Q0201-B13 and *CDFb-BN88:* The other two galaxies observed with ISAAC. As with SSA22a-MD41, we place lower limits on the star formation rate from Hα. Q0201-B13 shows some evidence of rotation in a slight tilt of the emission line, but the signal-to-noise ratio (S/N) is too low to construct a reasonable rotation curve.

Q1623-BX432, Q1623-BX449, Q1623-BX522, Q1623-MD107, Q1700-BX717, Q1700-MD109: These are the remaining objects in the sample. They span a factor of 3 in Hα luminosity, from Q1623-BX432 at the bright end to Q1623-BX449 at the faint end, but none show evidence of velocity shear. Our only kinematic information about these objects comes from the velocity dispersion; for three of the fainter objects (Q1623-BX449, Q1623-MD107, and Q1700-BX717) we were only able to place an upper limit on this quantity.

Non-detections: There are 10 galaxies that we observed with NIRSPEC but failed to detect. Four of these are accounted for by two observations in which we did not detect either of the galaxies we placed on the slit; in the cases of the other six, we detected one of the galaxies on the slit, but missed the other. For one of these the optical redshift was unknown, so our hopes for detecting it were not high. These 10 non-detections could have a variety of explanations, including errors in our optical redshifts (which are of marginal quality in many cases), in the astrometry, or in the guiding and tracking of the instrument and telescope. The objects could also be intrinsically faint due to extinction or a decline in the star formation rate, as discussed in § 2.5.

2.4 Kinematics

2.4.1 Rotation

Six galaxies in our sample of 16 show evidence of velocity shear, in the form of a spatially resolved, tilted Hα emission line. We have constructed rotation curves for these objects by fitting a Gaussian profile in wavelength to the emission line at each spatial location along the slit, summing three pixels in the spatial direction at each point in order to approximate the seeing of $\sim 0\rlap{.}''5$. Velocity offsets were measured with respect to the systemic redshift of the galaxy as determined from the central wavelength of the integrated Hα emission line; when possible, the spatial center was defined by summing the spectra in the dispersion direction without including the emission line and locating the center of the continuum. For those with no apparent continuum emission (Q1700-MD103 and Q1623-BX511), the center was defined as the spatial center of the emission line. The 2-D emission lines are shown in Figure 2.4 and the rotation curves in Figure 2.5. The observed velocities range from ~ 50 to ~ 240 km s^{-1}, comparable to those observed in local galaxies and up to $z \sim 1$ (Vogt et al. 1996, 1997). In most cases they show no sign of flattening at a terminal velocity; the blue-shifted end of the curve of West-BX600 is the only one that appears to flatten, and this is probably caused by imperfect subtraction of an adjacent sky line.

There are several systematic effects to be considered here; most of them result in an underestimation of the rotational velocity. Except in the case of SSA22a-MD41, no attempt was made to align the slit with the major axis of the galaxy (position angles were chosen in order to place two objects on the slit; see § 2.2); in fact, in most cases our ground-based images do not have sufficient resolution to allow the determination of a major axis. In the K'-band image of Q1700-BX691, however, it appears that here our slit was fortuitously aligned with the major axis of the galaxy. We also have an *HST* WFPC2 image of Q1623-BX447 (see Figure 2.3) in which it is apparent that the position angles of the slit and the galaxy differ by ~ 60 degrees. In the other three cases, the slit and the major axis were misaligned by an unknown amount. In addition the inclinations of the galaxies are not known. Given a random inclination and a random slit orientation, we will on average underestimate the rotational velocity by a factor of $(\pi/2)^2 \simeq 2.5$, where a factor of $\pi/2$ (the inverse of the average value of $\sin x$ over the interval $(0, \pi/2)$) comes from each effect. Also, because all or most of each galaxy falls within the slit, the velocity we measure at each spatial point along the slit is biased away from the maximum projected velocity at the major axis by the lower velocities of points away from the major axis. We must also consider the possibilities of uneven distribution of Hα emission and non-circular motions; both of these are likely, given the irregular morphologies of the galaxies (see Figure 2.3). A concentration of Hα away from the major axis of the galaxy would lead to an underestimate of the rotational velocity, but the effect of non-circular motions is more difficult to predict. Typically many of these effects are modeled and corrected for in rotation curves for less distant galaxies (Vogt et al. 1996, 1997; Swaters et al. 2003). Given the chaotic, or unknown, morphologies in our sample, we have not attempted to model these corrections.

We have used archival *HST* WFPC2 images that contain two of these galaxies, SSA22a-MD41 (in the F814W filter; proposal ID 5996) and Q1623-BX447 (F702W; proposal ID 6557). We reduced the images following the drizzling procedure outlined in the *HST* Dither Handbook (Koekemoer et al. 2002); see Fruchter & Hook (2002) for more details. The images are shown in Figure 2.3, with the position of the slit marked. Neither appears to be a well-formed disk; most of the rest-frame UV emission in SSA22a-MD41 is concentrated in a knot at the southwest edge, and Q1623-BX447 shows two distinct areas of emission. It is interesting to contrast these with images of two other galaxies for which we did not detect rotation: Figure 2.3 also shows images of Q1623-BX432 and Q1623-BX376, which are also contained in

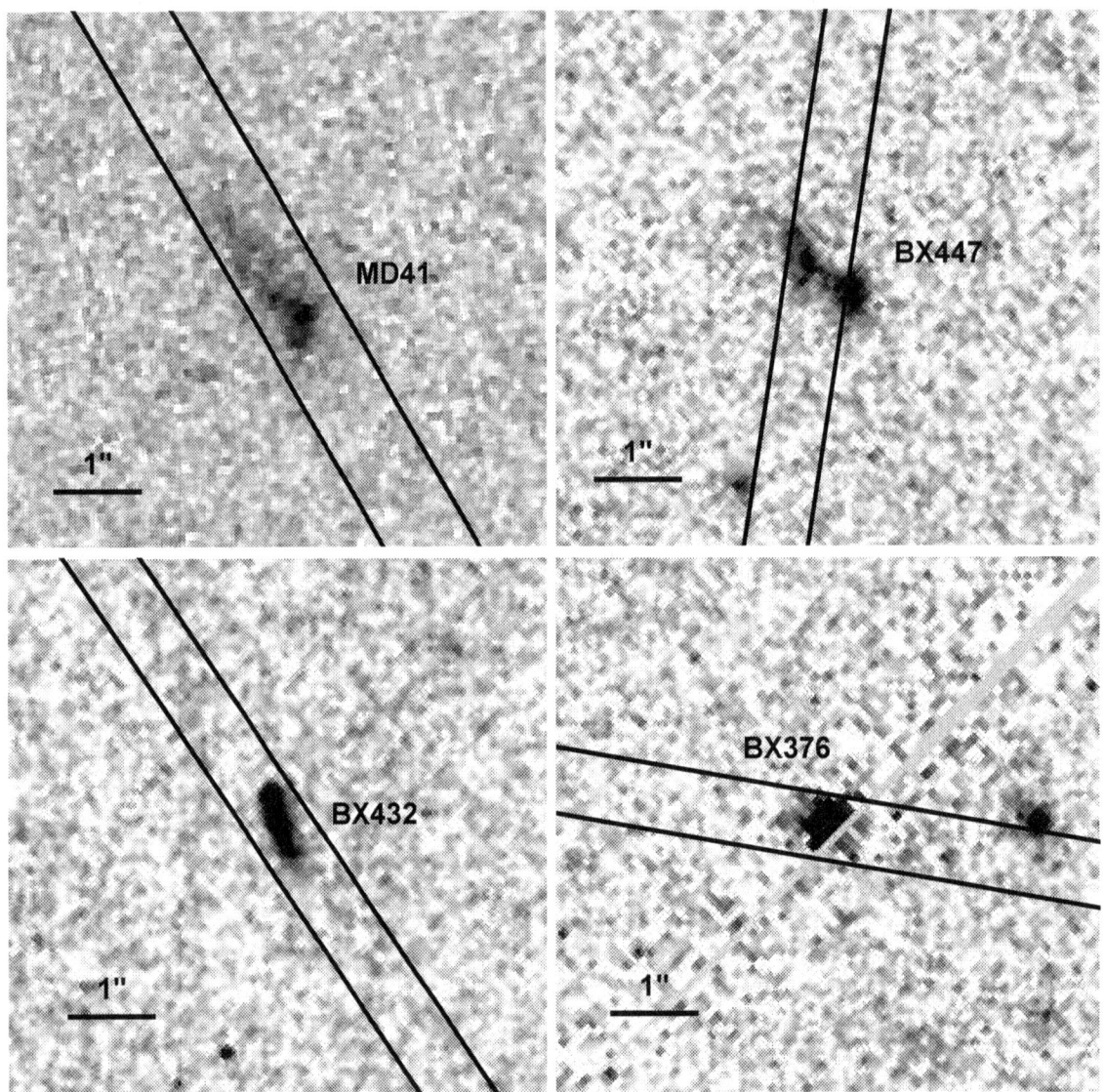

Figure 2.3 *HST* WFPC2 images of four of the galaxies in our sample. North is up and east to the left in all images, and positions of the slit are marked. *Upper left:* SSA22a-MD41, one of the objects in which we detect rotation, and the only one in which the slit was intentionally aligned with the major axis. *Upper right:* Q1623-BX447, another of the objects that show evidence of velocity shear. In this case the slit and the major axis were misaligned by ∼ 60 degrees. *Lower left:* We detect strong Hα emission from Q1623-BX432, but see no evidence of velocity shear despite the near alignment of the slit along the major axis. *Lower right:* Q1623-BX376, the object with the largest velocity dispersion in our sample. We also detect Hα emission at the same redshift from the object on the right; the two were classified as one extended object in our ground-based photometry. The gray line running through the image is the boundary between two of the wide-field detectors.

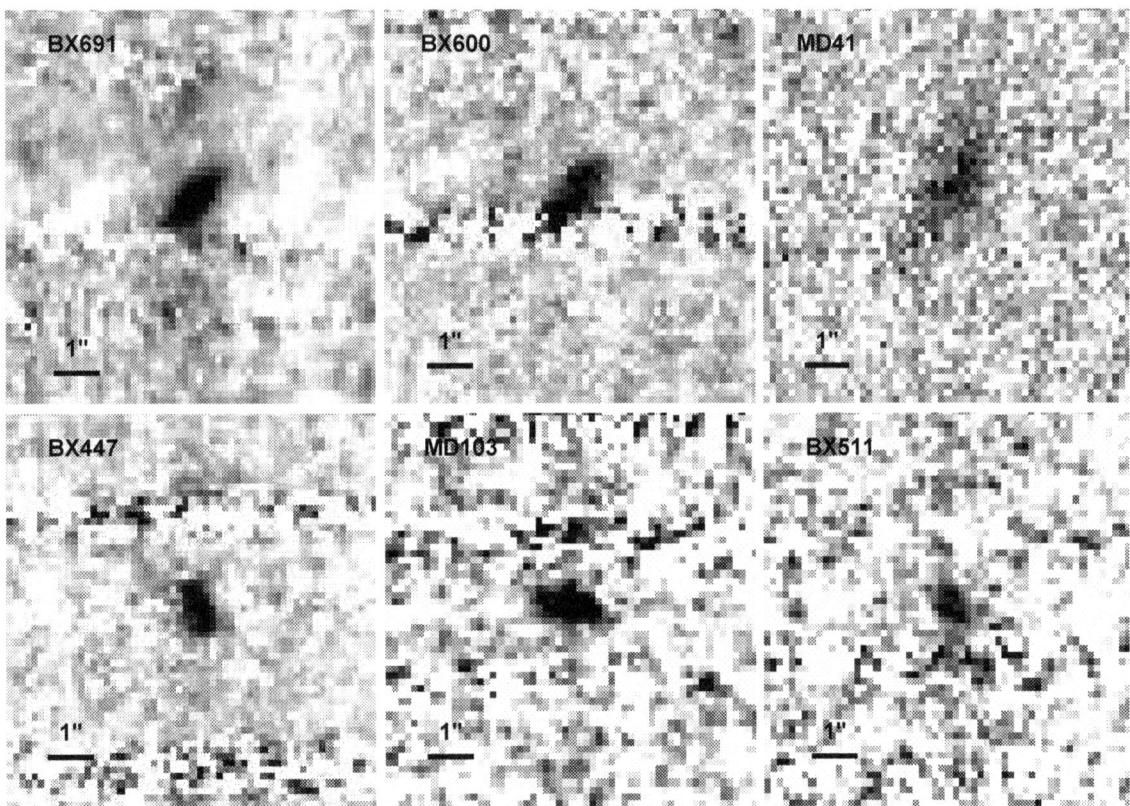

Figure 2.4 The two-dimensional spectra of the galaxies for which we have derived rotation curves, showing the tilt in the Hα emission line. From upper left, the galaxies are Q1700-BX691 at $z = 2.1895$; West-BX600 at $z = 2.1607$; SSA22a-MD41 at $z = 2.1713$; Q1623-BX447 at $z = 2.1481$; Q1700-MD103 at $z = 2.3148$; and Q1623-BX511 at $z = 2.2421$. A tilted [N II]$\lambda6584$ emission line is visible above Hα in the spectrum of Q1700-BX691. The x axis is spatial, with $1''$ scale bars shown, and y is the dispersion direction.

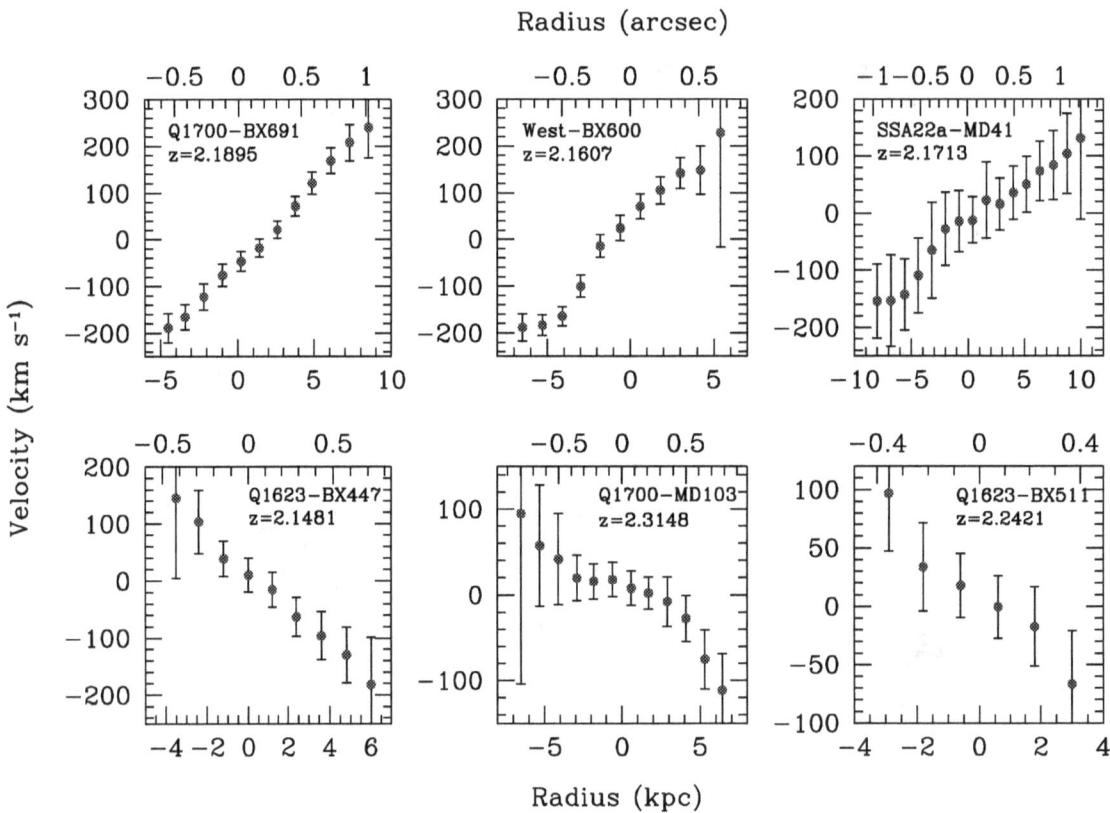

Figure 2.5 Rotation curves for the six galaxies that show spatially resolved, tilted Hα emission lines. From the rotational velocities and radii we derive lower limits on the mass of each galaxy; the mean dynamical mass is $\langle M_{dyn} \rangle \geq 4 \times 10^{10} M_\odot$. The galaxies are shown in order of decreasing mass; from upper left, Q1700-BX691, West-BX600, SSA22a-MD41, Q1623-BX447, Q1700-MD103, and Q1623-BX511. Note that the points are correlated due to the seeing of $\sim 5''$. We have used a cosmology with $H_0 = 70$ km s^{-1} Mpc^{-1}, $\Omega_m = 0.3$, and $\Omega_\Lambda = 0.7$ for the transformations between arcsec and kpc.

the Q1623 pointing and which also appear irregular. This demonstrates the difficulty of predicting the kinematics of these objects from even high-resolution imaging; complicated morphologies make inclinations and major axes difficult to determine, and objects with similar UV continuum morphologies may exhibit quite different Hα kinematic properties. We also point out that the Hα and UV emission may not be coincident; Pettini et al. (2001) observed nebular line emission extending ∼ 1″ beyond the UV emission in a galaxy at $z = 3.2$, and similar effects have been seen in local galaxies (Leitherer et al. 1996; de Mello et al. 1998; Johnson et al. 2000). Specifically, Conselice et al. (2000) compared Hα and UV emission in six nearby starburst galaxies, finding that the Hα and UV fluxes were well-correlated in three of the systems, but that they showed different morphologies in the other three.

Although we have no direct evidence that these galaxies are in fact disks, we make this assumption in order to use the radius r and the circular velocity v_c to calculate the enclosed mass,

$$M_{dyn} = v_c^2 r/G. \tag{2.3}$$

Since we have neither well-defined terminal velocities nor spatial centers for these objects, we have calculated lower limits on the masses by using half of the total spread in both velocity and distance, for v_c and r, respectively. We obtain an average dynamical mass of $\langle M \rangle \geq 4 \times 10^{10} M_\odot$; individual masses for each galaxy are shown in Table 2.2. As the Hα emission traces only the central star-forming regions of these objects, which are probably baryon-dominated, the masses derived are underestimates of the total halo masses of the galaxies. We can use an order of magnitude argument to estimate the total masses: for $\Omega_b = 0.02h^{-2}$ and $\Omega_m = 0.3$, $\Omega_m/\Omega_b \sim 7$ and the universe contains about six times more dark than baryonic matter. We therefore expect the total masses of the galaxies to be about seven times larger than their stellar masses, and we place a lower limit of $M \gtrsim 3 \times 10^{11} M_\odot$ on the typical halo mass of these galaxies. This is generally consistent with mass estimates from the clustering properties of LBGs at $z \sim 3$: Adelberger et al. (1998) find a typical mass of $8 \times 10^{11} h^{-1} M_\odot$ for a ΛCDM model, based on the number density and correlation length of the galaxies. Other analyses yield similar results (Baugh et al. 1998; Giavalisco & Dickinson 2001). We defer an analysis of the clustering of the $z \sim 2$ galaxies to a later work. We can also compare our mean baryonic mass with the median stellar mass from population synthesis models found for LBGs at $z \sim 3$ by Shapley et al. (2001), $m_{star} = 1.2 \times 10^{10} h^{-2} M_\odot$; again the two are in rough agreement.

There are few other examples of such rotation curves at redshifts of $z \gg 1$. Lemoine-Busserolle et al. (2003) have recently reported a rotation curve of a gravitationally lensed galaxy at $z = 1.9$; the rotation curve looks much like those we present here, with $v \gtrsim 200$ km s^{-1} at a radius of $\sim 1″$, although when the lensing correction is applied this radius corresponds to only ~ 1 kpc. Genzel et al. (2003) have used millimeter interferometry to observe rest-frame 335 μm continuum and CO(3–2) line emission from a massive submillimeter galaxy at $z = 2.8$; their data indicate a rotating disk with velocity ≥ 420 km s^{-1} at ~ 8 kpc in radius. From observations of [O III] at $z \simeq 3.2$, Moorwood et al. (2003) present a rotation curve with a velocity of 108 km s^{-1} at $\simeq 12$ kpc. Also in observations of [O III] and Hβ in 15 LBGs at $z \sim 3$, Pettini et al. (2001) see two spatially resolved and tilted emission lines, but the observed velocities reach only ~ 50 km s^{-1}. Simply counting the instances of rotation shows that the two samples are different at the 95% confidence level; the difference is actually more significant, because this test does not account for the larger rotational velocities at $z \sim 2$. It is interesting that we see stronger evidence for rotation in a sample of similar size at $z \sim 2$, and we will spend a moment speculating on the possible reasons for this. Poorer seeing during the $z \sim 3$ observations could perhaps account for the differences; this does not explain the larger values of the velocity dispersion σ we see at $z \sim 2$ (see § 2.4.2), however, as these should be unaffected even if the lines are spatially unresolved. It might then be that Hα is a more sensitive probe of rotation and velocity dispersion than [O III] because of higher surface brightness; but Pettini et al. (2001) typically measured [O III]λ5007/Hβ ~ 3, and Hα/Hβ ~ 3 as well, so Hα and [O III]λ5007 should have roughly comparable strengths in the $z \sim 3$ galaxies. We also note that in rotation curves for which they had both Hα and [O III]λ5007 data, Vogt et al. (1996) found that the flux distributions and velocities of the two lines matched well. Lemoine-Busserolle et al. (2003) also have both Hα and [O III]λ5007 observations for their rotation curve, and again the two lines give comparable results. The differences could also be due to S/N effects; but the $z \sim 3$ galaxies were generally observed with longer integration times than those in the current sample, and their spectra have S/N comparable to or higher than that of those presented here. We should also discuss the possibility that we may be observing different populations of galaxies at $z \sim 2$ and $z \sim 3$. We therefore consider the evidence for other intrinsic differences between galaxies at the two redshifts. The most obvious of these is apparent UV luminosity; the galaxies of Pettini et al. (2001) are brighter than those presented here, with only a few exceptions. This is simply because the brightest galaxies were selected for IR observation at $z \sim 3$, but not at $z \sim 2$. As discussed in § 2.2, however, the $z \sim 2$ selection criteria were chosen so that the galaxies they select would have SEDs similar to galaxies at $z \sim 3$. If we are indeed looking at different sets of objects at $z \sim 2$ and $z \sim 3$, both the average and range of their far-UV properties must be similar (although we do sample the luminosity function more deeply at $z \sim 2$). It is also possible that we are observing the two samples to different radii: Surface brightness is a strong function of redshift, scaling as $(1 + z)^4$, and this may limit the radii to which we can observe the galaxies at higher redshift. Star formation progressing to larger radii in the disks at later times could produce a similar effect. It is also possible that our stronger evidence for rotation reflects an increase in the number of rotating galaxies and their rotational speeds between $z \sim 3$ and $z \sim 2$. With the present data such a conclusion would be premature, however, since we cannot rule out all observational effects.

It is interesting to consider objects such as these in the context of hierarchical models of galaxy formation. We compare our data with predictions of the properties of LBGs at $z \sim 3$ (Mo, Mao, & White 1998, 1999), although it is not yet clear how the current sample and the $z \sim 3$ galaxies are related. LBGs are thought to be the central galaxies of the most massive dark halos present at $z \sim 3$, and they are predicted to be small and to have moderately high halo circular velocities but low stellar velocity dispersions. For a ΛCDM cosmology, Mo et al. (1999) predict that the median effective radius R_{eff} (defined as the semimajor axis of the isophote containing half of the star formation activity) is about $2\ h^{-1}$ kpc, and most galaxies should have R_{eff} between 0.8 and $5\ h^{-1}$ kpc. While the maximum radial extent of some of our rotation curves is larger than this, it is likely that the galaxies are visible at radii beyond R_{eff}, and these predictions are consistent with our measurements of half-light radii from the WFPC2 images. Mo et al. (1999) also predict a median halo circular velocity of 290 km s^{-1} for ΛCDM, with most galaxies falling between 220 and 400 km s^{-1}, and a median stellar velocity dispersion of ~ 120 km s^{-1}. Both of these predictions are reasonably consistent with our data, considering that we have not corrected our circular velocities for inclination or slit alignment effects, and that our velocities are lower limits due to the lack of flattening in the rotation curves. In fact, as noted above, the $z \sim 2$ galaxies are a better match to these predictions than the $z \sim 3$ LBGs, which have observed rotational velocities of only ~ 50–100 km s^{-1} and velocity dispersions of ~ 70 km s^{-1}.

Finally, additional observations will clarify the kinematics of the $z \sim 2$ sample. High resolution imaging in both the optical and the IR will allow a determination of the morphologies of the galaxies and the extent of the rest-frame optical emission; spectroscopic observations with varying position angles will provide strong constraints on rotating disk models. We are also optimistic about the possibility of obtaining a larger sample of rotation curves, since those presented here represent almost 40% of the galaxies observed. Looking farther into the future, integral field IR spectrographs that provide kinematic information at high spatial resolution over a contiguous region encompassing the entire galaxy will be ideal for probing the dynamics of high redshift galaxies; this may be the only way that the kinematic major axes of these objects can be determined.

2.4.2 Velocity Dispersions

We can obtain a limited amount of information about the dynamics and masses of the galaxies by simply measuring the widths of the emission lines. We have measured the one-dimensional velocity dispersion σ by fitting a Gaussian profile to each emission line, measuring its FWHM, and subtracting the instrumental broadening in quadrature from the FWHM. The instrumental broadening was measured from the widths of sky lines, and is ~ 15 Å for NIRSPEC and ~ 6 Å for ISAAC. The velocity dispersion is then the corrected FWHM divided by 2.355. We find a mean velocity dispersion of $\langle \sigma \rangle \sim 110$ km s^{-1}, with a maximum of 260 km s^{-1}. The dispersions for each galaxy are shown in Table 2.2, with 1 Δ_σ uncertainties from propagating the errors in each Gaussian fit (to avoid confusion stemming from overuse of the symbol σ, we use Δ_σ to represent the standard deviation in the velocity dispersion). Most of the lines are resolved; for those that are not we have set an upper limit of 2 Δ_σ. Our average velocity dispersion is $\sim 60\%$ higher than that found from the widths of [O III]$\lambda 5007$ and Hβ at $z \sim 3$ by Pettini et al. (2001), who found a median of ~ 70 km s^{-1}.[3]

Assuming that these velocities are due to motion of the gas in the gravitational potential of the galaxy, we can estimate the masses of the galaxies. For the simplified case of a uniform sphere,

$$M_{vir} = 5\sigma^2 (r_{1/2}/G). \tag{2.4}$$

From the *HST* image of the galaxies in the Q1623 field, we find $r_{1/2} \sim 0\overset{\prime\prime}{.}2$, which in our adopted cosmology corresponds to ~ 1.6 kpc at $z = 2.3$. We use this value to calculate the masses shown in Table 2.2. Accounting for the lower limits on four of the objects by using ASURV Rev. 1.2 (Lavalley, Isobe, & Feigelson 1992), a software package that calculates the statistical properties of samples containing limits or non-detections (survival analysis; Feigelson & Nelson 1985), we find a mean mass of $\sim 2 \times 10^{10} M_\odot$; this is in general agreement with the rotationally derived masses in § 2.4.1. As we noted when deriving masses from the rotation curves above, because the nebular emission comes mostly from the central star-forming regions of high-surface brightness, the velocity dispersions probably do not reflect the full gravitational potential of the galaxies.

There are several issues to consider in the interpretation of these mass estimates. In addition to the obvious caveats related to the assumption of spherical geometry, the uncertain value of $r_{1/2}$, and the sometimes large uncertainties in σ, we should consider whether or not the line broadening is indeed gravitational in origin. Galaxy-scale starburst-driven outflows with speeds of several hundred km s^{-1} have been shown to be ubiquitous in star-forming galaxies at $z \sim 3$ (Pettini et al. 2001). These are measured from the offsets of Lyα and the interstellar absorption lines with respect to the nebular emission lines taken to define the systemic velocity of the galaxy; Lyα is consistently redshifted with respect to the systemic velocity, while the interstellar lines are blueshifted. We are unable to determine conclusively whether or not similar outflows exist in the present sample, since in many cases the S/N ratios of our rest-frame UV spectra are too low to determine redshifts from Lyα and interstellar absorption lines with the necessary precision. However, for those objects that have spectra of sufficient quality, we have measured the velocities of the interstellar absorption lines and Lyα with respect to the Hα redshifts. The results are shown in Figure 2.6. We see that in this small sample, Lyα is consistently redshifted by several hundred km s^{-1}, but that the interstellar lines are both blueshifted and redshifted with respect to Hα. This offers marginal support for the existence of outflows, but clearly a larger sample is necessary. Even if these outflows do exist, however, it is not clear that they would result in an increase in the velocity dispersion. Our velocity dispersions are from Hα emission, which we take to be coming primarily from nebular gas at the systemic redshift of the galaxy, not from outflowing material. In addition, a correlation between the velocity dispersion and the

---

[3]We also find a mean of ~ 70 km s^{-1} in the [O III]$\lambda 5007$ velocity dispersions of a sample of 11 LBGs at $z \sim 3$, which we observed with NIRSPEC in April 2001. These data are unpublished, and will be described in detail in a later work.

speed of the outflow (here defined as the average of $v_{Ly\alpha} - v_{neb}$ and $v_{neb} - v_{IS}$) might be expected if the line broadening were due to outflowing gas. With this in mind we have examined a sample of 23 galaxies at $z \sim 3$ for which we have both velocity dispersions from the width of the [O III]λ5007 emission line and outflow velocities from the offsets between the nebular, interstellar absorption, and Lyα redshifts. We see no evidence for a strong link between the velocity dispersion and the speed of the outflow; the correlation coefficient between them is 0.13. These considerations lead us to believe that the presence of outflows is not a strong argument against gravitational broadening of the lines.

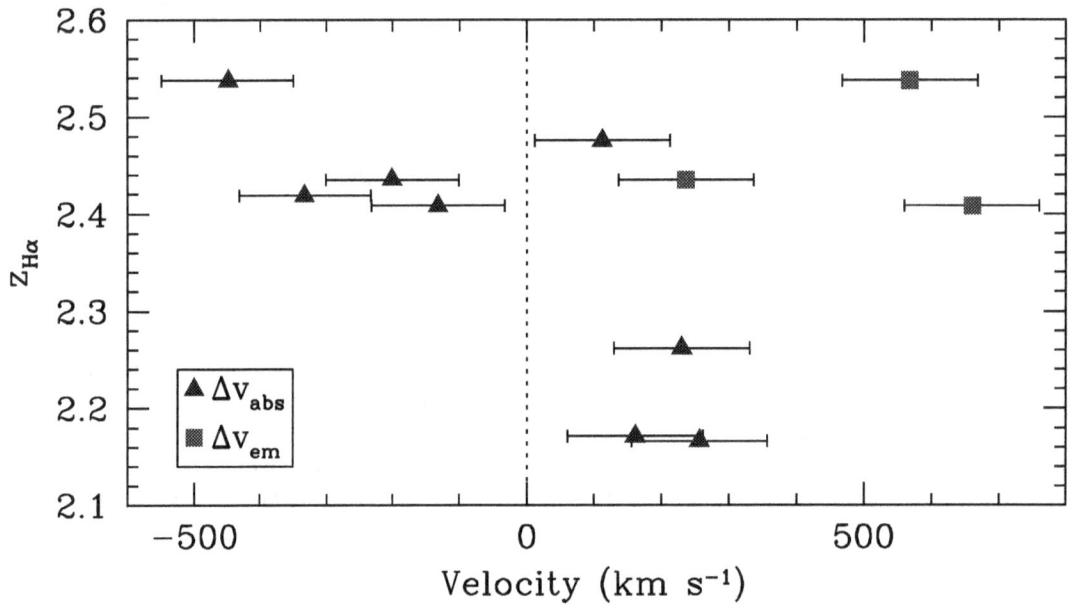

Figure 2.6 Velocity offsets between the systemic velocity of Hα and the velocities of the interstellar absorption lines (blue triangles) and Lyα emission (red squares). The sample is small because most of our galaxies do not have rest-frame UV spectra of sufficient S/N to make this comparison.

We are also struck by the spatial complexity of some of these objects. In particular, the Hα emission of Q1623-BX376 appears as two lines at the same redshift but separated by $2''.5$. The brighter of these, Q1623-BX376a, has the largest velocity dispersion in the sample, and shows an asymmetric line profile (see Figure 2.2), with a blueshifted tail extending about $0''.5$ in the opposite direction from the fainter component, Q1623-BX376b. It is primarily this tail that is responsible for the large velocity dispersion. This faint emission is also visible in the WFPC2 image shown in Figure 2.3 (where, unfortunately, the galaxy falls on the border between two of the wide-field detectors). Given the complicated structure of this object, we hesitate to attribute its broad emission line purely to random gravitational motions; galactic mergers or interactions could also produce such broadened emission lines and disturbed morphologies.

As a final test, we compare the one-dimensional velocity dispersion with luminosity. We see from Figure 2.7 that neither the 1500 Å continuum nor the Hα emission line luminosity correlates with velocity dispersion, either with or without a correction for extinction. Such a lack of correlation is also seen for galaxies at $z \sim 3$ (Pettini et al. 2001). This does not necessarily mean that the line widths are unrelated to the masses of the galaxies; it may be that large variations in the mass-to-light ratio are blurring any trend. We conclude that while these caveats are important, none of them provide a compelling argument against using the velocity dispersions to estimate the masses of the galaxies; therefore for the moment we will continue to do so.

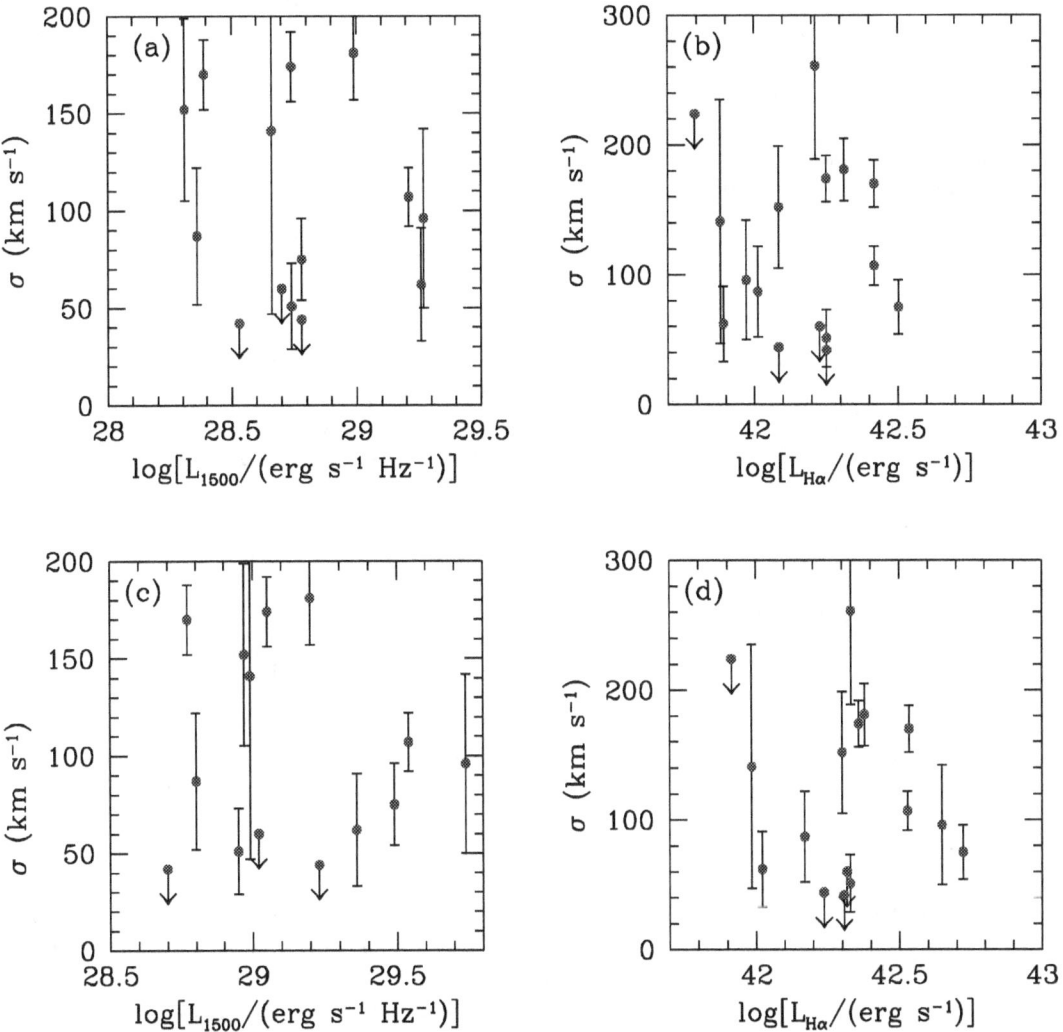

Figure 2.7 The velocity dispersion σ plotted against the 1500 Å continuum and Hα luminosities, without extinction corrections (a and b), and corrected as described in the text (c and d). Arrows indicate upper limits on σ. See § 2.5 for a discussion of the errors in luminosity.

2.5   Star Formation Rates and Extinction

Hα emission is one of the primary diagnostics of the star formation rate (SFR) in local galaxies, and therefore its observation at high redshift is particularly valuable for the sake of comparison with nearby samples. Redshifts of $z \lesssim 2.6$ are the highest at which Hα can currently be detected before it shifts out of the near-IR K-band window. Except for a few other observations of Hα at $z > 2$ (Teplitz et al. 1998; Kobulnicky & Koo 2000), most determinations of the star formation rate at high redshift have so far been based on the UV stellar continuum and, to a lesser extent, the Hβ emission line (Pettini et al. 2001). Here we compare star formation rates for the 16 galaxies in our sample deduced from the Hα flux and from the UV continuum emission; as the two are affected differently by dust and star formation history, our results can in principle tell us about the extinction and stellar populations of the galaxies. We have calculated Hα SFRs following Kennicutt (1998a):

$$\text{SFR } (\text{M}_\odot \text{ yr}^{-1}) = 7.9 \times 10^{-42} \text{ L(H}\alpha) \text{ (erg s}^{-1}). \tag{2.5}$$

The nebular recombination lines are a direct probe of the young, massive stellar population, since only the most massive and short-lived stars ($M \gtrsim 10\text{M}_\odot$) contribute significantly to the ionizing flux. Thus the emission lines provide a nearly instantaneous measure of the SFR, independent of the star formation history. The above equation assumes a Salpeter IMF with upper and lower mass cutoffs of 0.1 and 100 M$_\odot$ and case B recombination at T_e=10,000 K. It also assumes that all of the ionizing photons are reprocessed into nebular lines, i.e., that they are not absorbed by dust before they can ionize an atom, and that they do not escape the galaxy.

Ultraviolet-derived star formation rates were calculated from the broadband optical photometry, using the G magnitude as an approximation for the 1500 Å continuum (at $z = 2.3$, the mean redshift of our sample, the central wavelength of the G filter, 4830 Å, falls at a rest wavelength of 1464 Å). SFRs were calculated as follows (Kennicutt 1998a):

$$\text{SFR } (\text{M}_\odot \text{ yr}^{-1}) = 1.4 \times 10^{-28} \text{ L}_{1500} \text{ (erg s}^{-1} \text{ Hz}^{-1}). \tag{2.6}$$

This relationship applies to galaxies with continuous star formation over timescales of 10^8 years or longer; for a younger population, the UV continuum luminosity is still increasing as the number of massive stars increases, and the above equation will underestimate the star formation rate. The assumed IMF is the same as above.

The fluxes and corresponding SFRs are summarized in Table 2.3, and a comparison of the uncorrected star formation rates is shown in the left panel of Figure 2.8. The error bars reflect the uncertainties in flux calibration of the Hα emission and the UV photometry, about 25% and 10%, respectively; for the Hα spectra this includes both random and systematic error, as discussed in § 2.2.3, and is likely an underestimate in the noisiest cases. Uncertainties in the conversion from flux to SFR are not included. There are four objects for which we are only able to place lower limits on the SFR from Hα: Q1623-BX428, in which the Hα line fell on top of a strong sky line to which we have lost significant flux, and SSA22a-MD41, Q0201-B13, and CDFb-BN88, which were observed during non-photometric conditions (and calibrated with the least extinguished exposure of a standard, in order to place lower limits). Without correcting for extinction, we find SFR$_{H\alpha}$ > SFR$_{UV}$ in all but five cases; four of these are the lower limits described above. We find $\langle \text{SFR}_{H\alpha}/\text{SFR}_{UV} \rangle = 2.4$; this was computed using ASURV Rev. 1.2 (Lavalley, Isobe, & Feigelson 1992), a software package that calculates the statistical properties of samples containing limits or non-detections (survival analysis; Feigelson & Nelson 1985). This result is in qualitative agreement with previous observations of galaxies at $z \gtrsim 1$: Yan et al. (1999) find that the global star formation rate derived from Hα exceeds that from the UV by a factor of ~ 3, and Hopkins et al. (2000) obtain a measurement of SFR density from Hα at $0.7 \leq z \leq 1.8$ that is a factor of 2–3 greater than that estimated from UV data. Glazebrook et al.

(1999) study a sample of 13 galaxies at $z \sim 1$ from the Canada France Redshift Survey (CFRS); when the same Kennicutt (1998a) calibrations are used, their data give an Hα SFR 1.9 times higher than the UV SFR, without applying an extinction correction (Yan et al. 1999).[4] It is also comparable to the results of Bell & Kennicutt (2001), who find $\langle \mathrm{SFR_{H\alpha}/SFR_{UV}} \rangle = 1.5$ for galaxies with SFR $\gtrsim 1 \mathrm{M_\odot\ yr^{-1}}$ in a sample of 50 nearby star-forming galaxies. There is clearly a trend for the Hα-derived SFRs to be higher than those from the UV luminosity, in spite of differing selection criteria; both the Yan et al. (1999) and Hopkins et al. (2000) samples were selected in the IR, while ours is UV-selected and the Bell & Kennicutt (2001) sample is drawn from local galaxies observed by the Ultraviolet Imaging Telescope (UIT). We will discuss possible reasons for this trend below. We also note that the one remaining object with a larger UV SFR, Q1623-BX376, is a somewhat unusual case. It is bright and extended in the UV, and the Hα emission appears in two distinct lines at the same redshift but separated by $2''.5$. Since the UV photometry encompassed both components we have added the flux from both lines to calculate the Hα SFR, but it is clear from the WFPC2 image of Q1623-BX376 (Figure 2.3) that the fainter of the two components is largely off the edge of the slit; therefore we have likely missed some of the Hα emission.

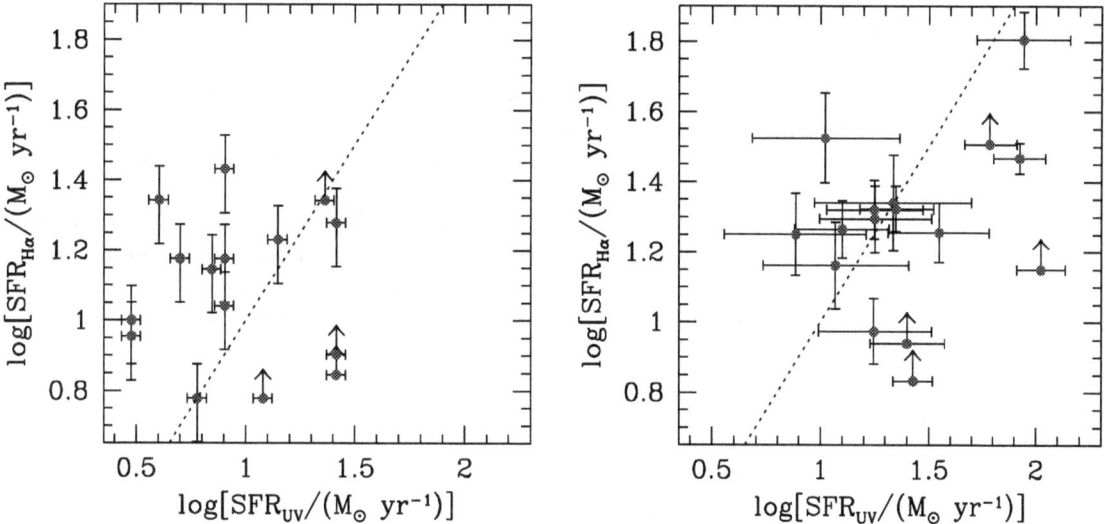

Figure 2.8 *Left:* Star formation rates from Hα and UV emission, uncorrected for extinction. Arrows indicate lower limits on the Hα SFR for objects observed during non-photometric conditions (SSA22a-MD41, Q0201-B13, and CDFb-BN88) or contaminated by sky lines (Q1623-BX428). Errors are 25% in $\mathrm{SFR_{H\alpha}}$ and 10% in $\mathrm{SFR_{UV}}$, reflecting uncertainties in flux calibration. Uncertainties in the conversion from flux to SFR are not included. *Right:* The SFRs corrected for extinction as described in § 2.5. The error bars reflect uncertainties in $E(B-V)$ only; flux calibration errors and errors in conversion from flux to SFR are not included. The dotted lines represent equal rates from Hα and UV emission.

There are at least two possible explanations for the larger Hα SFRs: dust extinction and the two star formation rate indicators' differing sensitivities to the ages of stellar populations and star formation histories. Our observations are consistent with the assumption that the ultraviolet emission generally suffers greater extinction than the Hα, as would be the case if both pass through the same clouds of dust.

---

[4]Yan et al. (1999) and Glazebrook et al. (1999) assume $H_0 = 50$ km s^{-1} Mpc^{-1} and $q_0 = 0.5$. Using this cosmology lowers our SFRs by 5–10%; the ratios of the Hα and UV rates are, of course, unaffected.

However, in analogy to local starbursts, it may be the case that the UV and nebular line emission come from different regions in the galaxies and encounter different amounts of dust accordingly (Calzetti 1997). In particular, it has been suggested that the most massive stars are still embedded in the dust clouds in which they formed, leading to greater extinction of the nebular line emission. This may be the case with Q1623-BX376, which is bright in the rest-frame UV, with Lyα emission and strong interstellar absorption lines, but which is undistinguished when observed in the rest-frame optical.

We can estimate the UV extinction using the observed broadband colors and an assumed spectral energy distribution (SED); we have calculated $E(B - V)$ in this way, using the $G - \mathcal{R}$ colors and an SED corresponding to continuous star formation with an age of 320 Myr, the median age found for LBGs at $z \sim 3$ by Shapley et al. (2001). Because extinction corrections are highly sensitive to errors in color measurements, we have made an effort to quantify the uncertainties and biases in our photometry. We added a large number of artificial galaxies of known colors and magnitudes to the actual images, and then recovered them using the same photometric tools that we applied to the real data (see Adelberger 2002; Steidel et al. 2003). We then selected artificial galaxies whose recovered colors match our selection criteria, and sorted them into bins by color and $\mathcal{R}$ magnitude. We used these to measure the mean and dispersion of $\Delta(G - \mathcal{R}) = ((G - \mathcal{R})_{\text{meas}} - (G - \mathcal{R})_{\text{true}})$, where the mean indicates systematic biases in the recovered colors and the dispersion reflects the characteristic measurement error, $\sigma(G - \mathcal{R})$. For the brightest galaxies in our sample ($\mathcal{R} < 23.5$) both of these quantities are small: $\langle\Delta(G - \mathcal{R})\rangle \simeq 0.03$ and $\sigma(G - \mathcal{R}) \simeq 0.05$. For those with $\mathcal{R} > 25$, we find $\langle\Delta(G - \mathcal{R})\rangle \simeq 0.04$ and $\sigma(G - \mathcal{R}) \simeq 0.14$. For each galaxy in our sample, we have used these statistics to correct the measured $G - \mathcal{R}$ color for the bias, and the color error has been propagated to determine uncertainties in $E(B - V)$; these range from 0.03 for the brightest galaxies to 0.08 for the faintest.

After calculating $E(B - V)$ in this way, we used the Calzetti et al. (2000) extinction law to correct the G magnitudes, and then used these to recalculate the UV star formation rates. For the sake of comparison we have also corrected the Hα fluxes, assuming the same values of $E(B - V)$; we found this to give better agreement between the corrected UV and Hα SFRs than the Calzetti (1997) relation $E_s(B - V) = (0.44 \pm 0.03)E_n(B - V)$ (where $E_s(B - V)$ is the color excess of the stellar continuum and $E_n(B - V)$ is that of the nebular emission lines). There may be some justification for this: If indeed there are galactic-scale outflows in these galaxies as in those at $z \sim 3$, then a screen of outflowing material may be obscuring all regions equally. Unfortunately we have no way of independently measuring the nebular extinction with our current data, as we do not have H-band measurements of Hβ. It should also be noted that the uncertainties inherent in flux calibration are too large to allow a reliable measurement of the Balmer decrement even if we had been able to obtain Hβ fluxes; for a Balmer decrement of 10%, expected for our mean $E(B - V) = 0.10$ mag, we would need to measure each line flux with an accuracy of 5% or less, far better than our current capabilities. The issue is further complicated by the fact that Hα and Hβ lie in different bands and cannot be observed simultaneously, so there may be a systematic offset between the flux calibrations of the two observations. It will therefore be difficult to test the Calzetti model directly.

A comparison of the extinction-corrected SFRs is shown in the right panel of Figure 2.8. They are in better agreement than the uncorrected SFRs, with $\langle\text{SFR}_{\text{Hα}}/\text{SFR}_{\text{UV}}\rangle = 1.2$ and a reduction in the scatter of 50% (1 σ; again accounting for the lower limits on four of the Hα SFRs). As emphasized above, the extinction correction is highly sensitive to uncertainties in the $G - \mathcal{R}$ colors; the errors bars reflect the errors in $E(B - V)$ determined above, propagated through to the star formation rates. Not shown are uncertainties in the extinction law, flux calibrations, or conversion of flux to star formation rate, all of which are considerable. Given these sources of error, and the uncertainty in the value of $E(B - V)$ that should be used for the nebular emission, the extinction-corrected SFRs should be taken with caution.

In Figure 2.9 we plot the ratio $\text{SFR}_{\text{neb}}/\text{SFR}_{\text{UV}}$ against the rest-frame UV continuum luminosity; none of these quantities have been corrected for extinction. We include data from Pettini et al. (2001), who used Hβ fluxes and the standard ratio Hα/Hβ = 2.75 (Osterbrock 1989) to calculate SFRs from recombination lines in galaxies at $z \sim 3$. We have also included unpublished data from our NIRSPEC run in April 2001; these are LBGs at $z \sim 3$, and star formation rates have been calculated in the same way as in Pettini et al. (2001). These data will be discussed in detail in a future paper. The dotted curves represent lines of constant nebular line SFR, and the number at the top of each curve is its SFR_{neb}, in $M_\odot$ yr^{-1}. We see that there is a moderate trend for the UV-faint galaxies to have higher nebular line SFRs relative to their UV SFRs, as might be the case if these objects were more heavily reddened. From the curves of constant SFR_{neb} it can be seen that galaxies with similar nebular line luminosities and varying amounts of UV extinction will naturally follow such a trend. As noted in § 2.3, the UV luminosities of the galaxies in our sample vary by a factor of 9, while the Hα luminosities are the same to within a factor of 4; this is consistent with the idea that the galaxies in our sample have roughly the same SFR, but differ in the amount of UV extinction. This model may offer an explanation for the difference between our results and those of Pettini et al. (2001), who observed no tendency for $SFR_{H\beta}$ to be systematically greater than SFR_{UV}. As is apparent from the figure, the galaxies in their sample are brighter in the UV than all but four of those presented here, and could plausibly suffer less extinction. Several caveats are in order, however. We observe no correlation between either the $(G-\mathcal{R})$ color or $E(B-V)$ and the ratio of SFRs; if reddening is indeed the cause of the observed trend, then UV continuum measurements are not sufficient to quantify it. We also note that many objects with faint Hα emission would fall in the lower left corner of the plot; this is apparent when we add the objects we failed to detect to the figure (shown as magenta stars). We have plotted only those objects that were placed on the slit with another galaxy that was detected, so that we know our astrometry was correct. We have placed upper limits on their Hα star formation rates by assigning a maximum $SFR_{H\alpha}$ corresponding to 1 σ less than the flux of our weakest detection, and we have calculated UV SFRs based on their photometry as with the rest of the sample. It is clear from this exercise that the absence of data points in the lower left is a selection effect; such galaxies would have undetectably small SFRs. The absence of data points in the upper right is more significant, as these objects would be easily detectable; from the curves of constant SFR_{neb}, we see that any galaxies falling here would have extremely large SFRs. In spite of these cautions, we believe that this figure is consistent with a model in which reddening is the primary cause of the discrepancy between the two SFR indicators.

Changes in the star formation rate on short timescales could also be reflected in our differing star formation rates, since Hα emission is a more instantaneous measure of the SFR than the UV emission. The nebular recombination lines are the reprocessed light of only the most massive ($M \gtrsim 10 M_\odot$) and short-lived stars, while the UV emission probes a wider mass range ($M \gtrsim 5 M_\odot$). Therefore a starburst that has begun in the past $\sim 10^8$ yr will not yet have reached full UV luminosity and will have an underestimated UV star formation rate, whereas a decline in star formation will cause an immediate decrease in Hα emission as the most massive stars die off. In a large sample of galaxies with redshifts $0 < z < 0.4$, Sullivan et al. (2000) find that the UV flux indicates a consistently higher SFR than the Hα, and that the discrepancy is best explained by short bursts of star formation superimposed on a smooth star formation history. Such a model could also explain the larger UV SFR of a galaxy such as Q1623-BX376; however, this relationship between the UV and Hα SFRs is strongest at the fainter end of the Sullivan et al. (2000) sample, whereas Q1623-BX376 would fall at the bright end. As noted above, there were several galaxies that we observed but failed to detect. This could be explained by a decline in the star formation rate, but due to the difficulties presented by the sky background in the IR, the marginal quality of some of our optical redshifts, and the possibility of errors in astrometry or the guiding and tracking of the instrument and telescope, these objects have not been included in the statistical comparison of SFRs.

In the following paragraphs we explain why we believe that a young stellar population is not the primary cause of the discrepancy between the SFRs. As we have no information on the ages of the stellar populations of the galaxies in our sample, we will assume that they are similar to LBGs at $z \sim 3$, although as we have pointed out above, the samples at $z \sim 2$ and $z \sim 3$ have different kinematic properties and the $z \sim 3$ sample tends to cover brighter UV luminosities. The stellar populations of LBGs at $z \sim 3$

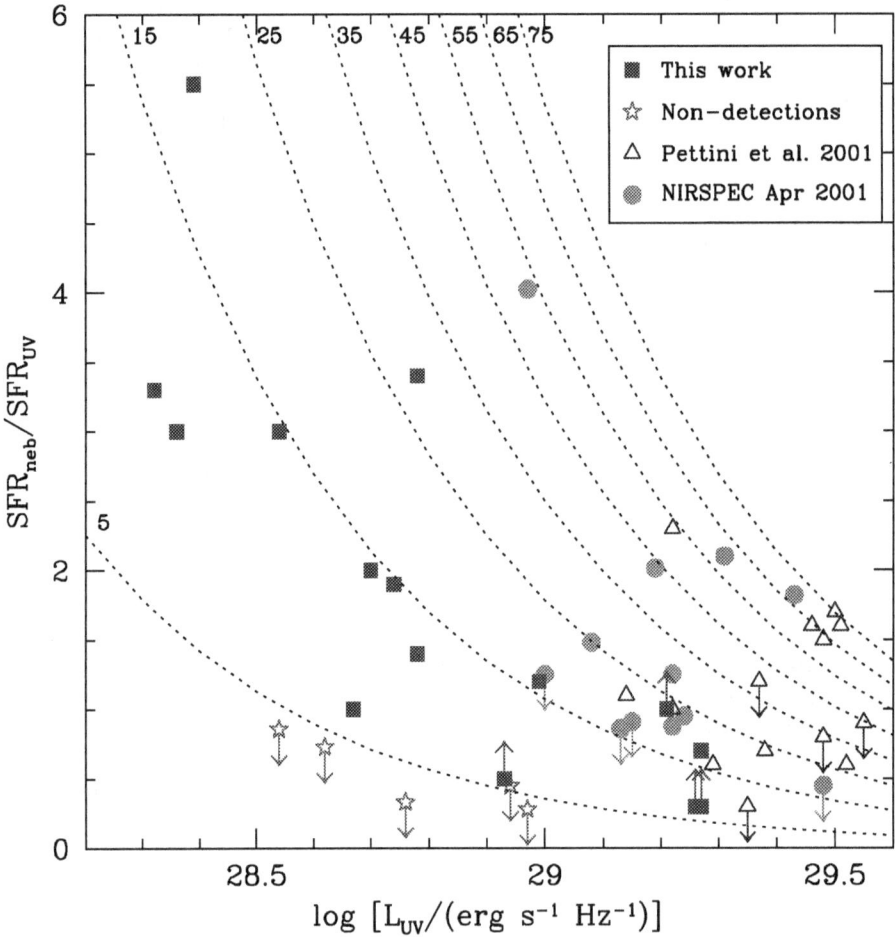

Figure 2.9 A comparison of the ratio of SFRs from nebular line and UV continuum emission with UV continuum luminosity. Red squares are the current data set, magenta stars represent upper limits on the ratio of SFRs for the five objects that we did not detect while successfully observing another galaxy on the same slit, blue triangles are from the $z \sim 3$ sample of Pettini et al. (2001), and green circles are unpublished $z \sim 3$ data from our April 2001 NIRSPEC run. The dotted curves are lines of constant $\mathrm{SFR_{neb}}$, and the number at the top of each curve is its $\mathrm{SFR_{neb}}$ in $\mathrm{M_\odot\ yr^{-1}}$. UV luminosity is computed from the G magnitude for the $z \sim 2$ sample; the center of the G filter corresponds to ~ 1500 Å at $z = 2.3$. For the $z \sim 3$ sample we use the $\mathcal{R}$ magnitude, corresponding to ~ 1700 Å at $z = 3$. SFRs for the Pettini et al. (2001) and April 2001 samples were calculated from Hβ emission, assuming Hα/H$\beta = 2.75$ and applying the Kennicutt (1998a) conversion from Hα to SFR. Errors are suppressed for clarity, but are $\sim 25\%$ in $\mathrm{SFR_{H\alpha}}$ and 10% in $\mathrm{SFR_{UV}}$ and G as discussed in the text. See Pettini et al. (2001) for discussion of errors in the SFRs from Hβ.

are now well-studied (Shapley et al. 2001; Papovich et al. 2001), and the Papovich et al. (2001) sample includes some galaxies in the range $z = 2$–2.5. Population synthesis models for a sample of 81 LBGs by Shapley et al. (2001) give a median age since the onset of the most recent episode of star formation of $t_{\rm sf} \simeq 320$ Myr, with more than 40% having $t_{\rm sf} > 500$ Myr and 25% having $t_{\rm sf} < 40$ Myr. We might then expect $\sim 25\%$ of our sample to have an underestimated UV SFR; however, the youngest galaxies in the Shapley et al. (2001) sample are also the most extinguished and have the highest star formation rates. Among those with $t_{\rm sf} < 100$ Myr, the mean $E(B - V)$ is 0.27, higher than that of any of the objects in our sample and $3\,\sigma$ higher than our sample mean of 0.10. The mean star formation rate among the same subset is 261 $M_\odot$ yr^{-1}, far higher than that of any of the objects in our sample even after correcting for extinction. Assuming that star-forming galaxies at $z \sim 2$ are similar to those at $z \sim 3$, it is therefore unlikely that the stellar populations of our sample are young enough to account for the difference in SFRs.

Papovich et al. (2001) fit a set of detailed models to 33 LBGs in the Hubble Deep Field North, finding that the age distribution is strongly dependent on metallicity, IMF, the choice of extinction law, and the assumed star formation history. It is possible to vary these parameters to make the ages young enough to lead to an underestimate of the UV star formation rate; the youngest ages, $\langle t \rangle \simeq 40$ Myr, are given by a Scalo IMF with 0.2 $Z_\odot$. Although this may be a reasonable estimate for the metallicity of these objects— Pettini et al. (2001) find 0.1 to 0.5 $Z_\odot$ for galaxies at $z \sim 3$—the theoretical stellar atmospheres used in the population synthesis models are not well-tested for low metallicities, and the results should therefore be treated with caution. More generally, even ages as young as these cannot fully explain the discrepancy between the SFRs. The mean factor of 2.4 difference between the Hα and UV rates would require the average UV luminosity to have reached only $\sim 40\%$ of its full value, which occurs less than 5 Myr after the beginning of a burst of continuous star formation. Such an extremely young age is unphysical; the time required for a burst of star formation to propagate across a galaxy is approximately the dynamical timescale, and $t_{dyn} \simeq 30$ Myr for galaxies of the masses and sizes found in § 2.4. We can state the timescale argument in another way as well: The average stellar mass of our galaxies, $\langle M \rangle \gtrsim 4 \times 10^{10} M_\odot$, combined with an assumed age of 2 Gyr, gives a characteristic $\dot{M} \sim 20 M_\odot$ yr^{-1}, about the same as our mean Hα SFR of 16 $M_\odot$ yr^{-1}. This implies that the current SFRs of the galaxies are similar to their past averages over the last 2 Gyr, and that a current burst is unlikely. Assuming an age younger than 2 Gyr, a mass larger than our lower limit of $4 \times 10^{10} M_\odot$, or significant gas recycling results in a current SFR less than the past average, excluding a current burst even further.

The effects of dust and star formation history are indistinguishable in individual cases; in the sample taken as a whole, the systematic depression of SFR$_{\rm UV}$ relative to SFR$_{\rm H\alpha}$ suggests that extinction is the dominant effect, since variations in star formation history would induce scatter in the plots rather than systematic effects. Our knowledge of star formation and extinction at high redshift generally supports this conclusion. A moderate amount of extinction is indicated by our data, with a mean $E(B - V)$ of 0.10 (corresponding to $A_{1500} \sim 1$ mag and attenuation by a factor of ~ 2.5, using the Calzetti et al. (2000) extinction law); in studies of LBGs at $z \sim 3$, Shapley et al. (2001) find a median dust attenuation factor of ~ 4.5 at ~ 1500 Å, while Papovich et al. (2001) find a factor of 3.0–4.4, depending on metallicity. Our results also provide some support for previous estimates of UV extinction at high redshift: If the Hα extinction is assumed to be about the same as it is in local galaxies, a typical factor of 2, and if we assume that the factor of ~ 2.4 reduction in SFR$_{\rm UV}$ relative to SFR$_{\rm H\alpha}$ is due to extinction, then we obtain a UV extinction factor of ~ 5, the same as that applied to the UV luminosity density at $z \sim 3$ by Steidel et al. (1999). We also note that this is in general agreement with the average UV attenuation factor of 5–6 obtained from studies of the X-ray luminosity of LBGs at $z \sim 3$ (Nandra et al. 2002). In summary, while we cannot rule out the effects of star formation history entirely, our results are consistent with other estimates of extinction in galaxies at high redshift, and such extinction naturally explains the differences we see in the Hα- and UV-derived star formation rates.

2.6   Summary and Conclusions

We have presented Hα spectroscopy of 16 galaxies in the redshift range $2.0 < z < 2.6$; this is so far the largest sample of near-IR spectra of galaxies at these redshifts. The galaxies were selected based on their broadband rest-frame UV colors, using an adaptation of the technique used to select Lyman break galaxies at $z \sim 3$. Those observed here are drawn from a large sample of such galaxies, with redshifts already confirmed; because proximity to a QSO sightline was the primary selection criterion for near-IR observation, we believe the 16 galaxies presented here to be representative of the sample as a whole. We have analyzed the spectra in order to determine the kinematic and star-forming properties of the galaxies, and we reach the following conclusions:

1. Six of the 16 galaxies show spatially extended, tilted Hα emission lines, such as would be produced by ordered rotation. Rotation curves for these galaxies show a mean velocity of ~ 150 km s^{-1} at a mean radius of ~ 6 kpc; these are lower limits obtained by taking half of the total range in both velocity and distance. Measuring from the spatial location of the continuum and the dynamical center of the lines, we obtain a maximum velocity of ~ 240 km s^{-1} and a maximum radius of 10 kpc in the most extreme cases. We have obtained archival *HST* images for two of these galaxies, and they appear to be morphologically irregular, as do all of the other galaxies in our sample for which we have such images. Because of their chaotic morphologies, we have not attempted to model any corrections to the rotation curves. We have used the lower limits on the rotational velocity and radius of each galaxy to derive a dynamical mass; we obtain a mean of $\langle M \rangle \geq 4 \times 10^{10} M_\odot$. Because Hα emission probes only the central star-forming regions of the galaxies, we expect their total halo masses to be several times larger. These results are in general agreement with the predictions of models of hierarchical galaxy formation for LBGs at $z \sim 3$.

2. Values of the one-dimensional velocity dispersion σ range from 50 to 260 km s^{-1}, with a mean of ~ 110 km s^{-1}. Assuming that the line widths are due to gravitational motions in the potentials of the galaxies, the mean virial mass implied is $2 \times 10^{10} M_\odot$; this is in general agreement with the masses we obtain from the rotation curves. We consider other possible origins for the broadening of the lines, including large-scale outflows, mergers and interactions.

3. Both the rotational velocity v_c and the velocity dispersion σ tend to be larger at $z \sim 2$ than at $z \sim 3$. We see evidence of rotation in $\sim 40\%$ of our sample, whereas Pettini et al. (2001) found such evidence in only $\sim 10\%$ of a sample of similar size at $z \sim 3$. Furthermore, we find rotational velocities of ~ 150 km s^{-1}, as compared to ~ 50 km s^{-1} at $z \sim 3$. Our mean value of σ, ~ 110 km s^{-1}, is $\sim 60\%$ larger than the value found at $z \sim 3$ by Pettini et al. (2001). We have considered possible selection effects that may explain these systematic differences, but have not found a convincing explanation. It may be that the redshift dependence of surface brightness allows us to sample to larger radii at $z \sim 2$, or that our photometric selection criteria pick out different populations of galaxies at $z \sim 2$ and $z \sim 3$. It is also possible that the effect is real and reflects the growth of disks between these two epochs.

4. We use the Hα luminosity to calculate the star formation rates of the galaxies, and compare these to the SFRs derived from the rest-frame UV continuum luminosity. We use the calibrations of Kennicutt (1998a) in both cases. We obtain a mean SFR$_{\text{Hα}}$ of 16 M$_\odot$ yr^{-1}, and a mean SFR$_{\text{Hα}}$/SFR$_{\text{UV}}$ ratio of 2.4. After correcting both luminosities for extinction using the Calzetti et al. (2000) extinction law, we find SFR$_{\text{Hα}}$/SFR$_{\text{UV}} = 1.2$, with a 50% reduction in scatter. We discuss the effects of extinction and star formation history on the SFRs, and conclude that extinction is the more likely explanation for their discrepancy. We also see a moderate correlation between the ratio SFR$_{\text{Hα}}$/SFR$_{\text{UV}}$ and the UV luminosities of the galaxies, such that UV-faint galaxies have a higher SFR$_{\text{Hα}}$/SFR$_{\text{UV}}$ ratio. Such an effect could be produced if the fainter galaxies undergo more extinction.

5. Finally, we expect that many of the points discussed here will become clearer as the sample of near-IR observations of galaxies at these redshifts grows. The photometric technique for selecting galaxies at $z \sim 2$ has so far produced hundreds of galaxies with confirmed redshifts in this range, and further

observations of their kinematics, line fluxes, and morphologies will shed light on star formation, extinction, and the formation of disks at high redshift.

We would like to thank the referee for many helpful comments and suggestions. We also thank the staffs at the Keck and VLT observatories for their competent assistance with the observations. CCS, DKE, AES, and MPH have been supported by grant AST-0070773 from the U.S. National Science Foundation and by the David and Lucile Packard Foundation. KLA acknowledges support from the Harvard Society of Fellows. Finally, we wish to extend special thanks to those of Hawaiian ancestry on whose sacred mountain we are privileged to be guests. Without their generous hospitality, most of the observations presented herein would not have been possible.

Table 2.1. Galaxies Observed

Galaxy	R.A. (J2000)	Dec. (J2000)	$z_{Ly\alpha}$[a]	z_{abs}[b]	$z_{H\alpha}$[c]	$\mathcal{R}$	$G - \mathcal{R}$	Exposure (s)	Telescope/Instrument
CDFb-BN88	00:53:52.87	12:23:51.25	—	2.263	2.2615	23.14	0.29	12×720	VLT 1/ISAAC
Q0201-B13	02:03:49.25	11:36:10.58	—	2.167	2.1663	23.34	0.02	16×720	VLT 1/ISAAC
Westphal-BX600[d]	14:17:15.55	52:36:15.64	—	—	2.1607	23.94	0.10	5×900	Keck II/NIRSPEC
Q1623-BX376	16:25:05.63	26:46:49.12	2.415	2.408	2.4085	23.31	0.24	4×900	Keck II/NIRSPEC
Q1623-BX428	16:25:48.42	26:47:40.24	—	2.053	2.0538[e]	23.95	0.13	4×900	Keck II/NIRSPEC
Q1623-BX432	16:25:48.74	26:46:47.05	2.187	2.180	2.1817	24.58	0.10	4×900	Keck II/NIRSPEC
Q1623-BX447	16:25:50.38	26:47:14.07	—	2.149	2.1481	24.48	0.17	4×900	Keck II/NIRSPEC
Q1623-BX449	16:25:50.55	26:46:59.63	—	2.417	2.4188	24.86	0.20	4×900	Keck II/NIRSPEC
Q1623-BX511	16:25:56.10	26:44:44.38	—	2.246	2.2421[e]	25.37	0.42	4×900	Keck II/NIRSPEC
Q1623-BX522	16:25:55.76	26:44:53.17	—	2.476	2.4757	24.50	0.31	4×900	Keck II/NIRSPEC
Q1623-MD107	16:25:53.88	26:45:15.19	2.543	2.536	2.5373	25.35	0.12	4×900	Keck II/NIRSPEC
Q1700-BX691	17:01:05.99	64:12:10.27	—	2.189	2.1895	25.33	0.22	4×900	Keck II/NIRSPEC
Q1700-BX717	17:00:57.00	64:12:23.71	2.438	—	2.4353	24.78	0.20	4×900	Keck II/NIRSPEC
Q1700-MD103	17:01:00.20	64:11:56.00	—	2.308	2.3148	24.23	0.46	900+600	Keck II/NIRSPEC
Q1700-MD109	17:01:04.48	64:12:09.28	2.295	2.297	2.2942	25.46	0.26	4×900	Keck II/NIRSPEC
SSA22a-MD41	22:17:39.97	00:17:11.04	—	2.173	2.1713	23.31	0.19	15×720	VLT 1/ISAAC

[a] Vacuum heliocentric redshift of Lyα emission line, when present.

[b] Vacuum heliocentric redshift from rest-frame UV interstellar absorption lines.

[c] Vacuum heliocentric redshift of Hα emission line.

[d] We have not yet obtained a rest-frame UV spectrum of Westphal-BX600

[e] The Hα redshifts of the galaxies Q1623-BX428 and Q1623-BX511 are somewhat uncertain due to the presence of strong sky lines near Hα.

Table 2.2.   Kinematics

Galaxy	$z_{H\alpha}$[a]	σ (km s^{-1})	v_c[b] (km s^{-1})	M_{vir}[c] (M$_\odot$)	M_{dyn}[d] (M$_\odot$)
CDFb-BN88	2.2615	96 ± 46	—	1.7×10^{10}	—
Q0201-B13	2.1663	62 ± 29	—	7.2×10^{9}	—
Westphal-BX600	2.1607	181 ± 24	~ 210	6.2×10^{10}	6.0×10^{10}
Q1623-BX376a	2.4085	261 ± 72	—	1.3×10^{11}	—
Q1623-BX376b	2.4085	< 224	—	$< 9.4 \times 10^{10}$	—
Q1623-BX428[e]	2.0538	—	—	—	—
Q1623-BX432	2.1817	51 ± 22	—	5.0×10^{9}	—
Q1623-BX447	2.1481	174 ± 18	~ 160	5.8×10^{10}	3.0×10^{10}
Q1623-BX449	2.4188	141 ± 94	—	3.7×10^{10}	—
Q1623-BX511	2.2421	152 ± 47	~ 80	4.4×10^{10}	4.6×10^{9}
Q1623-BX522	2.4757	< 44	—	$< 3.8 \times 10^{10}$	—
Q1623-MD107	2.5373	< 42	—	$< 2.9 \times 10^{10}$	—
Q1700-BX691	2.1895	170 ± 18	~ 220	5.5×10^{10}	7.0×10^{10}
Q1700-BX717	2.4353	< 60	—	$< 1.3 \times 10^{10}$	—
Q1700-MD103	2.3148	75 ± 21	~ 100	1.1×10^{10}	1.6×10^{10}
Q1700-MD109	2.2942	87 ± 35	—	1.5×10^{10}	—
SSA22a-MD41	2.1713	107 ± 15	~ 150	2.2×10^{10}	4.2×10^{10}

[a]Vacuum heliocentric redshift of Hα emission line.

[b]Minimum rotational velocity, $(v_{max} - v_{min})/2$.

[c]Masses calculated from the velocity dispersion.

[d]Minimum masses derived from rotational velocities when available.

[e]Sky line contamination prevented a measurement of σ.

Table 2.3.   Fluxes and Star Formation Rates

Galaxy	$z_{H\alpha}$ [a]	$\mathcal{R}$	$G-\mathcal{R}$	$F_{H\alpha}$ [b]	$L_{H\alpha}$ [c]	$E(B-V)$ [d]	$SFR_{H\alpha}$ [e]	$SFR_{H\alpha}$ [f]	SFR_{UV} [g]	SFR_{UV} [h]	$\frac{SFR_{H\alpha}}{SFR_{UV}}$ [i]
CDFb-BN88	2.2615	23.14	0.29	2.6	1.0	0.146	>8	>14	26 ± 3	106^{+31}_{-24}	>0.3
Q0201-B13	2.1663	23.34	0.02	2.4	0.8	0.004	>7	>7	26 ± 3	27^{+6}_{-5}	>0.3
Westphal-BX600	2.1607	23.94	0.10	6.3	2.2	0.048	17 ± 4	21^{+3}_{-3}	14 ± 1	22^{+11}_{-7}	1.2
Q1623-BX376	2.4085	23.31	0.24	5.3	2.4	0.111	19 ± 5	29^{+3}_{-3}	26 ± 3	84^{+27}_{-20}	0.7
Q1623-BX428	2.0538	23.95	0.13	2.7	0.8	0.073	>6	>9	12 ± 1	25^{+12}_{-8}	>0.5
Q1623-BX432	2.1817	24.58	0.10	5.4	1.9	0.048	15 ± 4	18^{+4}_{-3}	8 ± 1	13^{+8}_{-5}	1.9
Q1623-BX447	2.1481	24.48	0.17	5.6	1.9	0.082	15 ± 4	21^{+5}_{-4}	8 ± 1	18^{+12}_{-7}	1.9
Q1623-BX449	2.4188	24.86	0.20	1.8	0.8	0.094	6 ± 2	9^{+2}_{-2}	6 ± 1	18^{+15}_{-8}	1.0
Q1623-BX511	2.2421	25.37	0.42	3.4	1.3	0.194	10 ± 3	22^{+6}_{-5}	3 ± 0.3	22^{+28}_{-12}	3.3
Q1623-BX522	2.4757	24.50	0.31	2.8	1.3	0.132	11 ± 3	18^{+4}_{-3}	8 ± 1	35^{+26}_{-15}	1.4
Q1623-MD107	2.5373	25.35	0.12	3.7	1.9	0.043	15 ± 4	18^{+6}_{-4}	5 ± 1	8^{+9}_{-4}	3.0
Q1700-BX691	2.1895	25.33	0.22	7.7	2.8	0.108	22 ± 6	33^{+12}_{-9}	4 ± 0.4	10^{+12}_{-7}	5.5

Table 2.3—Continued

Galaxy	$z_{\mathrm{H}\alpha}$[a]	$\mathcal{R}$	$G-\mathcal{R}$	$F_{\mathrm{H}\alpha}$[b]	$L_{\mathrm{H}\alpha}$[c]	$E(B-V)$[d]	$\mathrm{SFR}_{\mathrm{H}\alpha}$[e]	$\mathrm{SFR}_{\mathrm{H}\alpha}$[f]	$\mathrm{SFR}_{\mathrm{UV}}$[g]	$\mathrm{SFR}_{\mathrm{UV}}$[h]	$\frac{\mathrm{SFR}_{\mathrm{H}\alpha}}{\mathrm{SFR}_{\mathrm{UV}}}$[i]
Q1700-BX717	2.4353	24.78	0.20	3.8	1.8	0.087	14 ± 4	20^{+5}_{-4}	7 ± 1	18^{+15}_{-8}	2.0
Q1700-MD103	2.3148	24.23	0.46	8.2	3.4	0.224	27 ± 7	64^{+13}_{-11}	8 ± 1	88^{+56}_{-35}	3.4
Q1700-MD109	2.2942	25.46	0.26	2.8	1.1	0.124	9 ± 2	14^{+5}_{-4}	3 ± 0.3	12^{+14}_{-6}	3.0
SSA22a-MD41	2.1713	23.31	0.19	7.9	2.8	0.097	>22	>32	23 ± 2	61^{+20}_{-15}	>1.0
Mean value[j]	2.2787	24.37	0.21	4.6	1.8	0.101	16	26	12	35	2.4

[a]Vacuum heliocentric redshift of Hα line.

[b]Line flux in units of 10^{-17} erg s^{-1} cm^{-2}.

[c]Luminosity in units of 10^{42} erg s^{-1}.

[d]From $G-\mathcal{R}$ colors, corrected as described in § 2.5.

[e]SFR in M$_\odot$ yr^{-1} from Hα luminosity, uncorrected for extinction.

[f]SFR in M$_\odot$ yr^{-1} from Hα luminosity, corrected for extinction.

[g]SFR in M$_\odot$ yr^{-1} from G magnitude, uncorrected for extinction.

[h]SFR in M$_\odot$ yr^{-1} from G magnitude, corrected for extinction.

[i]Ratio of uncorrected SFRs.

[j]For those quantities containing lower limits, statistics are computed using survival analysis as discussed in § 2.5.

Note. — $H_0 = 70$ km s^{-1} Mpc^{-1}, $\Omega_m = 0.3$, and $\Omega_\Lambda = 0.7$.

Chapter 3

The Kinematics of Morphologically Selected $z \sim 2$ Galaxies in the GOODS-North Field[*][†]

Dawn K. Erb,[a] Charles C. Steidel,[a] Alice E. Shapley,[b] Max Pettini,[c] Kurt L. Adelberger[d]

[a]California Institute of Technology, MS 105–24, Pasadena, CA 91125

[b]Department of Astronomy, 601 Campbell Hall, University of California at Berkeley, Berkeley, CA 94720

[c]Institute of Astronomy, Madingley Road, Cambridge CB3 0HA, UK

[d]Carnegie Observatories, 813 Santa Barbara Street, Pasadena, CA 91101

Abstract

We present near-IR spectra of Hα emission from 13 galaxies at $z \sim 2$ in the GOODS-N field. The galaxies were selected primarily because they appear to have elongated morphologies, and slits were aligned with the major axes (as determined from the rest-frame UV emission) of 11 of the 13. If the galaxies are elongated because they are highly inclined, alignment of the slit and major axis should maximize the observed velocity and reveal velocity shear, if present. In spite of this alignment, we see spatially resolved velocity shear in only two galaxies. We show that the seeing makes a large difference in the observed velocity spread of a tilted emission line, and use this information to place limits on the velocity spread of the ionized gas of the galaxies in the sample: We find that all 13 have $v_{0.5} \leq 110$ km s^{-1}, where $v_{0.5}$ is the velocity shear (half of the velocity range of a tilted emission line) that would be observed under our best seeing conditions of $\sim 0\farcs5$. When combined with previous work, our data also indicate that aligning the slit along the major axis does not increase the probability of observing a tilted emission line. We then focus on the one-dimensional velocity dispersion σ, which is much less affected by the seeing, and see that the elongated subsample exhibits a significantly *lower* velocity dispersion than galaxies selected at random from our total Hα sample, not higher as one might have expected. We also see some evidence that the elongated galaxies are less reddened than those randomly selected using only UV colors. Both of these results are counter to what would be expected if the elongated galaxies were highly inclined disks. It is at least as likely that the galaxies' elongated morphologies are due to merging subunits.

---

[*]Based on data obtained at the W.M. Keck Observatory, which is operated as a scientific partnership among the California Institute of Technology, the University of California, and NASA, and was made possible by the generous financial support of the W.M. Keck Foundation.

[†]A version of this chapter was published in *The Astrophysical Journal*, vol. 612, 122–130.

3.1   Introduction

At redshifts up to $z \sim 1$, it is possible to identify disk galaxies, place spectroscopic slits along the galaxies' major axes, and obtain rotation curves similar to those of disk galaxies in the local universe (e.g., Vogt et al. 1996, 1997; Böhm et al. 2004). As redshift increases, galaxy morphologies become increasingly irregular (van den Bergh 2001; Conselice et al. 2004), and spatially resolved spectra for kinematic measurements are more difficult to obtain as galaxy sizes approach the size of the seeing disk and the width of slits. In addition, the rest-frame optical emission lines used for such measurements shift into the near-IR for $z \gtrsim 1.4$. Nevertheless spatially resolved and tilted rest-frame optical emission lines from which "rotation curves" can be constructed have been seen for galaxies beyond $z \sim 1.5$ (Pettini et al. 2001; Lemoine-Busserolle et al. 2003; Moorwood et al. 2003; Erb et al. 2003); these lines generally have low signal-to-noise ratios (S/N) and the morphologies of the galaxies are irregular or unknown, making the interpretation of the tilted lines uncertain. One-dimensional velocity dispersions that do not require spatial resolution are much easier to obtain at high redshift; their use as a mass indicator for such galaxies has been well-studied (e.g., Kobulnicky & Gebhardt 2000), although faint emission lines may not trace a galaxy's full gravitational potential. Whether or not an emission line is spatially resolved, the alignment of a slit with a galaxy's apparent major axis should maximize the observed velocity, if the galaxy is elongated because it is highly inclined. Such a slit orientation also tests the alignment of the morphological and kinematic major axes.

Observations such as these at $z \sim 2$ are critical, as it has become increasingly clear that this epoch is an important period in the evolution of the universe. Galaxies at $z \sim 3$ are compact, rapidly star-forming, and morphogically disordered (e.g., Giavalisco et al. 1996; Shapley et al. 2001), whereas those at $z \lesssim 1$ have become the normal Hubble sequence galaxies of the universe today. Recent results suggest that QSO activity reaches a peak near $z \sim 2$ (Fan et al. 2001), and the median redshift of bright submillimeter galaxies is $z = 2.4$ (Chapman et al. 2003). In addition, most of the stellar mass in the universe today formed during this epoch: In a study of galaxies in the HDF-N, Dickinson et al. (2003) find that 50–75% of the mass in today's galaxies had formed by $z \sim 1$, but only 3–14% had formed by $z \approx 2.7$.

In this paper we present kinematic measurements from Hα emission in galaxies at $z \sim 2$ in the GOODS-N field, making use of the deep *HST* ACS imaging that has recently been obtained as part of the Great Observatories Origins Deep Survey (GOODS; Giavalisco et al. 2004). We describe our selection criteria, observations, and data reduction in §3.2; the results of our simple morphological analyses in §3.3; and our kinematic results in §3.4. In §3.4.1 we highlight the critical importance of the seeing to kinematic measurements at high redshift, and we discuss our results in §3.5. We use a cosmology with $H_0 = 70$ km s^{-1} Mpc^{-1}, $\Omega_m = 0.3$, and $\Omega_\Lambda = 0.7$ throughout. In such a cosmology, $1''$ corresponds to 8.1 kpc at $z = 2.38$, the mean redshift of the current sample.

3.2   Observations and Data Reduction

The 13 galaxies presented here lie in the Hubble Deep Field North region imaged with the *HST* Advanced Camera for Surveys (ACS) as part of the GOODS program. We have spectroscopically identified approximately 180 galaxies at $z \sim 2$ in this field, using $U_n G \mathcal{R}$ color selection criteria and rest-frame UV spectra from the LRIS-B spectrograh on the 10 m W.M. Keck I telescope on Mauna Kea (Adelberger et al. 2005a; Steidel et al. 2004). We have observed a subset of 13 of these with the near-infrared spectrograph NIR-SPEC (McLean et al. 1998), on the W.M. Keck II telescope. The galaxies were selected primarily because they appeared elongated on our initial inspection of the z-band GOODS-N data (detailed morphological analysis was not done until the full V1.0 release of the GOODS data in late August 2003). We also tried to select galaxies with redshifts that put Hα in a favorable position with respect to the night sky lines, and chose two (BX305, BX1368) for this reason, as well as for their somewhat more compact appearance for the sake of comparison. BX1085 and BX1086 were observed because it was possible to place both of them on the slit with the elongated galaxy BX1084, as discussed below. The final column in Table 3.1 summarizes the selection criterion for each galaxy.

We attempted to align the slit with the elongated axis in all cases, with the exception of BX1084, BX1085 and BX1086[1], which lie on a single line and were therefore observed simultaneously with the PA set by their relative orientation. Coincidentally this PA differs from the PA of BX1084 by only 19°. Because morphological analysis was done after the observations, the slit and galaxy PAs were slightly misaligned, with an average offset of 10°. Eleven out of 13 galaxies therefore had slits aligned to within 25° of the galaxy PA, and 9 out of 13 to within 13°. Such slight misalignments will not prevent the detection of significant rotation, should it exist (Vogt et al. 1996; Simard & Pritchet 1998; Böhm et al. 2004; Metevier et al. 2004). It is also possible for misalignment of the slit and the galaxy to introduce the appearance of velocity shear where none may actually exist; in §3.4.1 below we explain why this effect has not biased the current observations. The galaxy PAs and slit positions are given in Table 3.2.

Most of the observations were conducted on the nights of 7–9 May 2003 (UT), with two additional objects (BX1055 and BX1397) observed on 7–8 July 2003 (UT). We used the $0\rlap{.}''76 \times 42''$ slit for all observations. The seeing was relatively poor during the May run, with FWHM $0\rlap{.}''7$–$0\rlap{.}''9$ in the K-band; in July the seeing was better, with FWHM $\sim 0\rlap{.}''5$. Conditions were not photometric on either run, and in particular much of the May data suffers from significant losses due to cirrus clouds. For a detailed description of the observing and data reduction procedures, see Erb et al. (2003).

3.3 Morphologies

HST ACS images of the 13 galaxies in our sample are shown in Figure 3.1. For the purpose of morphological analysis we combined the four ACS bands into a single image, weighting in order to maximize the S/N. The images shown therefore cover the approximate range 1300–2500 Å in the rest frame. The slits used are marked with heavy blue lines, and the object PA with fine red lines. As noted in §3.2, the average misalignment of slit and galaxy is 10°; the slit and galaxy PAs are aligned to within 25° for 11 of 13 objects, and to within 13° for 9 of 13. Most of the galaxies appear irregular on simple inspection, and we have not attempted any kind of morphological classification. We have however performed some simple morphological analysis, with two goals in mind: to determine whether there is any correlation between a galaxy's kinematic properties and its aspect ratio, and to determine a size to be used in mass estimates.

Morphological analysis of faint galaxies is difficult, as it requires the separation of low surface brightness galaxy pixels from the sky background. We have estimated galaxy shapes and sizes using the pixels that make up half of the sky-subtracted light within a $1\rlap{.}''5$ (30 pixel) radius around the object centroid. The centroid is determined using an iterative process that calculates the centroid of an object, subtracts the sky value determined from a surrounding annulus, and recomputes the centroid until convergence is reached. We then use the pixels within $1\rlap{.}''5$ of the final centroid for the remaining analysis. We calculate two measures of galaxy size as follows: The effective half-light radius $r_{1/2} = (A/\pi)^{1/2}$ (where A is the area of pixels encompassing 50% of the galaxy's light) is sensitive to how large the bright regions of an object are, but not to their distribution; i.e., it depends on the number of pixels required to make up half of the galaxy's light, but not on how those pixels are distributed within the 30 pixel radius. We find a mean and standard deviation of $\langle r_{1/2} \rangle = 0\rlap{.}''24 \pm 0\rlap{.}''06$, corresponding to $\langle r_{1/2} \rangle = 1.9 \pm 0.5$ kpc. We also calculate d_{maj}, the RMS dispersion of the light about the centroid along the major axis (we avoid using the symbol σ to

---

[1]BX1086 is the only galaxy in the sample for which we did not previously know the redshift. We believe that the detected line is Hα because the interloper fraction from the BX color selection criteria is $\sim 9\%$, and the typical interlopers (star-forming galaxies at $\langle z \rangle = 0.17$) do not have a single strong emission line in the K-band. The Hα redshift is also nearly identical to the redshift of BX1084, which is separated from BX1086 by $2\rlap{.}''9$.

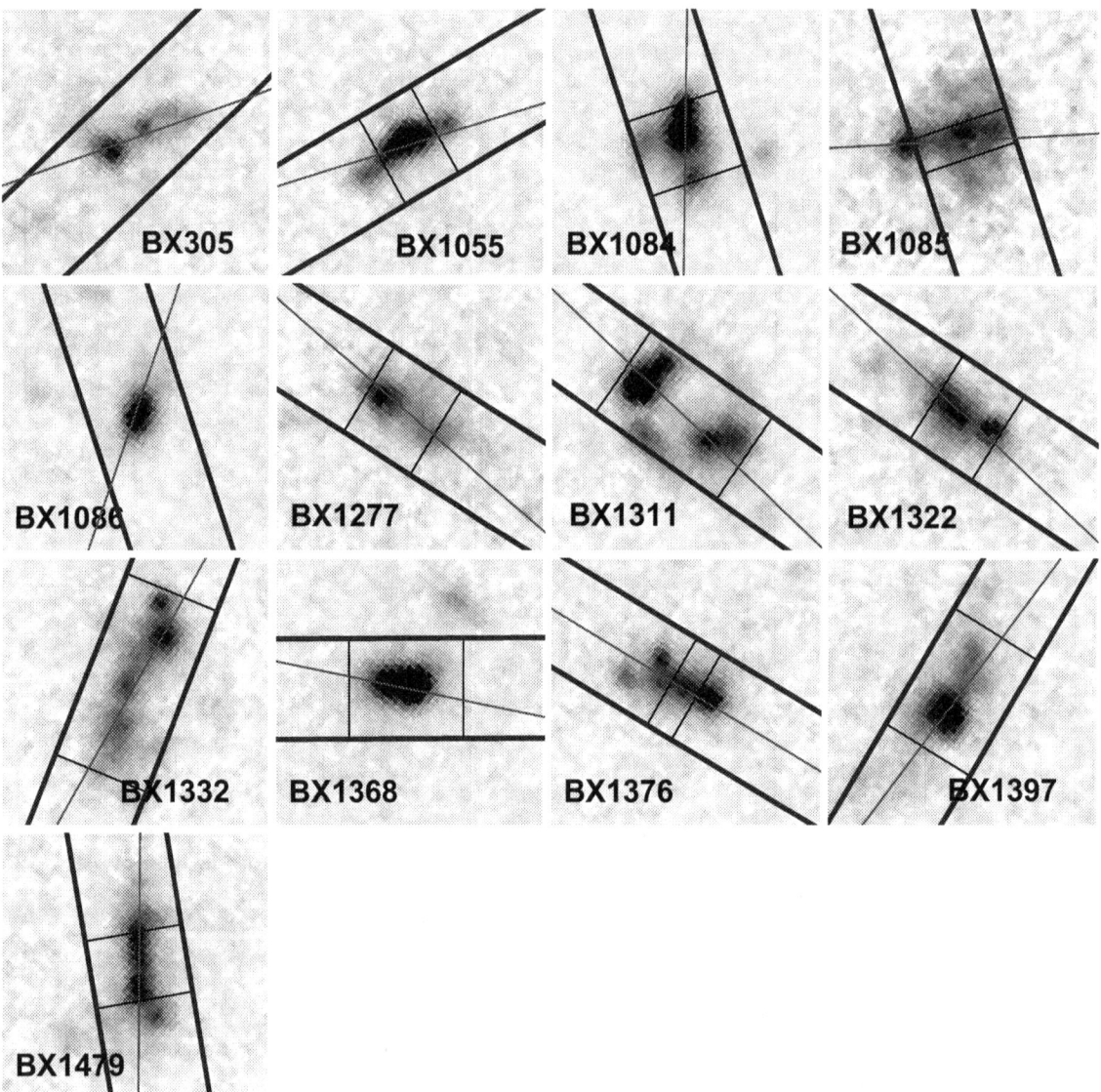

Figure 3.1 HST ACS images of the 13 galaxies in our sample, from the GOODS program. The $BViz$ bands have been combined to maximize S/N, and the images shown cover the approximate range of 1300–2500 Å in the rest frame. The 0″.76-wide slit (6.2 kpc at $z = 2.3$) is marked in blue, and the red line shows the galaxy PA. The thin lines perpendicular to the slit show the approximate extent of the Hα emission after deconvolution of the seeing. No lines are shown for BX305 and BX1086 because they appeared to be point sources after deconvolution. All images are oriented with N up and E to the left, and the pixel scale is 0″.05/pixel.

eliminate confusion with velocity dispersions calculated below), and the aspect ratio $a = d_{min}/d_{maj}$, the ratio of the dispersions along the minor and major axes. We find $\langle d_{maj} \rangle = 0\rlap{.}''20 \pm 0\rlap{.}''09$, or $\langle d_{maj} \rangle = 1.7 \pm 0.7$ kpc, and $\langle a \rangle = 0.41 \pm 0.12$. We use d_{maj} for mass estimates. Individual values of these parameters are shown in Table 3.2, and in the following section we relate them to the galaxies' kinematic properties.

3.4   Kinematic Results

Close examination of the 13 galaxies' two-dimensional spectra reveals only two showing spatially resolved velocity shear. We show velocity curves for the two emission lines in Figure 3.2, constructed by fitting a Gaussian profile in wavelength at each spatial position along the emission line, summing three pixels to increase the S/N. We refer to these figures as velocity curves rather than rotation curves because it is not clear that the tilted lines are caused by rotation; see §3.5 for discussion of this issue. The first of the tilted emission lines, that of BX1332, extends almost $1\rlap{.}''75$ in the spatial direction; this is significantly larger than the $\sim 0\rlap{.}''8$ seeing disk, and therefore the modest tilt of ~ 90 km s^{-1} peak-to-peak is readily apparent. Assuming for the moment that this tilt is due to rotation, we calculate a lower limit on the circular velocity v_c of half of the total velocity spread, $v_c \sim 45$ km s^{-1}. We emphasize here that we use $v_c \equiv (v_{max} - v_{min})/2$ as an observed quantity defined as half of the velocity spread of the emission line; this is not the terminal circular velocity, and is almost certainly less than that velocity. The second galaxy, BX1397, which was observed in July 2003 when the seeing was $0\rlap{.}''5$, has a smaller spatial extent of $\sim 1.2''$ but a larger velocity range, with $v_c \sim 110$ km s^{-1}. The large spatial extent of BX1332 may be suggested by the fact that it is the most elongated of the 13 galaxies, with $a = 0.21$; but BX1397 has $a = 0.47$, slightly higher than the mean $\langle a \rangle = 0.41$. Clearly the aspect ratio does not predict rotation. Such a low incidence of rotation in galaxies with slits placed along their elongated axes is perhaps surprising, especially since our earlier near-IR spectroscopy found tilted lines in about 40% of the galaxies observed, with random (with one exception) slit orientations (Erb et al. 2003). In §3.4.1 below we discuss limitations on observed rotational velocities imposed by the seeing, and in §3.5 we consider the implications of our low incidence of observed rotation.

It is also interesting to compare the relative extent of the Hα and rest-frame UV emission. BX1332 and BX1397 have values of d_{maj} of $0\rlap{.}''37$ and $0\rlap{.}''20$ respectively, but the total spatial extent of all the pixels considered to be part of the galaxy for morphological analysis is $\sim 1\rlap{.}''5$ for BX1332 and $\sim 0\rlap{.}''9$ for BX1397. Deconvolving the seeing from the Hα sizes above, we find $\sim 1\rlap{.}''6$ for BX1332 and $\sim 1\rlap{.}''1$ for BX1397, values that agree well with the total UV sizes. We can make this comparison for the rest of the objects by measuring the spatial width of the Hα emission line and again deconvolving the seeing. The results of this calculation are shown by the thin lines perpendicular to the slits in Figure 3.1, which show the approximate extent of the Hα emission. No lines are shown for BX305 and BX1086 because they appeared to be point sources after deconvolution. In general the UV and Hα emission agree well, although there are cases with greater extents of both UV (BX1376) and Hα (BX1368). The spatial extent of the Hα emission line of each galaxy is given in Table 3.2.

We have calculated one-dimensional velocity dispersions for the galaxies in the sample by fitting a Gaussian profile to each Hα emission line using the splot task in IRAF, which also provides errors. We deconvolved the instrumental profile by subtracting the instrumental FWHM of 15 Å (measured from the widths of sky lines) in quadrature from the Gaussian FWHM; the instrumental profile is comparable to the widths of the lines, so this deconvolution is important. It is possible to obtain higher resolution if the object and the seeing are smaller than the slit, but because the seeing was comparable to the slit width for most of our observations we have used the instrumental resolution as measured from the sky lines. After converting the FWHM to a velocity, we computed the velocity dispersion $\sigma = $ FWHM/2.355. The results are shown in column 8 of Table 3.2, with errors of one standard deviation from propagating the error in

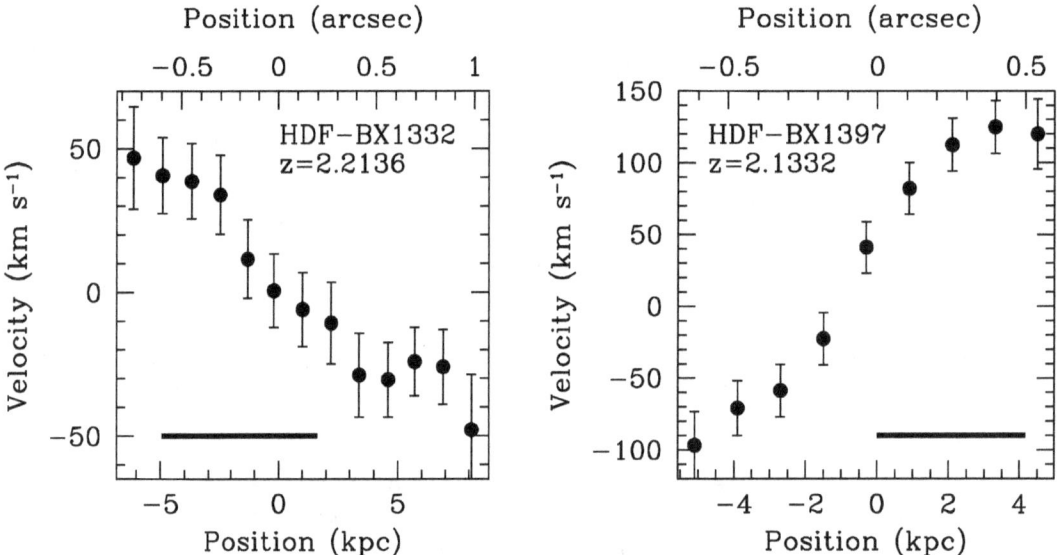

Figure 3.2 Velocity curves for BX1332 and BX1397, the only two galaxies in the sample that show tilted emission lines. The horizontal black lines represent the size of the seeing disk. The points are strongly correlated due to the seeing, but the spatial extent of each galaxy is larger than the seeing disk and the two endpoints used to determine v_c are not correlated.

the Gaussian line fit. Two of the lines (BX1055 and BX1322) had observed widths less than 10% greater than the instrumental profile, making the deconvolution highly uncertain; for these lines we place an upper limit on σ of twice the one-standard-deviation error. The measured width of BX1085 was slightly less than the instrumental FWHM, preventing a calculation of σ entirely. Neglecting the objects with upper limits, we find a mean velocity dispersion of $\langle\sigma\rangle = 92$ km s^{-1}, with a standard deviation of 34 km s^{-1}; this is somewhat smaller than the mean of the total sample of 61 galaxies at $z \sim 2$ for which we currently have Hα measurements[2], $\langle\sigma\rangle \sim 114 \pm 51$ km s^{-1}. We have combined these measurements of σ with our estimates of galaxy sizes d_{maj} to determine masses. We calculate the virial mass $M_{\mathrm{vir}} = 5\sigma^2(d_{\mathrm{maj}}/G)$, and find a mean and standard deviation of $\langle M_{\mathrm{vir}}\rangle = (1.6 \pm 1.1) \times 10^{10} \mathrm{M}_\odot$. The results of the mass calculations are shown in Table 3.2.

3.4.1 Seeing

3.4.1.1 Tilted Emission Lines

There is a long list of factors that affect the observed rotation curves of disk galaxies; these include the alignment of the slit with the major axis of the galaxy, the width of the slit compared to the size of the galaxy, the inclination of the disk, the spatial and velocity distribution of emission from the galaxy, and the blurring effect of the seeing (see, e.g., Vogt et al. 1996, 1997; Somerville & Primack 1999; Böhm et al. 2004; Metevier et al. 2004 for discussion of these effects). We concentrate here on the seeing, which can have a tremendous effect on the velocities observed in galaxies at high redshift. We can see this in two different observations of the galaxy Q1700-BX691 ($z = 2.1895$), which we first observed with NIRSPEC in May 2002 (Erb et al. 2003); during these observations the seeing was $\sim 0\rlap{.}''5$. Using the identical instrumental configuration and position angle on the sky, we observed the galaxy again in May 2003, when the seeing was $\sim 0\rlap{.}''9$. In Figure 3.3 we compare velocity curves derived from the two observations: At left is the previously published velocity curve, which has $v_c \sim 220$ km s^{-1}, and on the right is a velocity curve from the May 2003 observations, constructed in identical fashion and plotted on the same scale for comparison. It covers approximately the same spatial range as the previous observation, but shows about half the range in velocity, with $v_c \sim 120$ km s^{-1}. It also shows some distortion in the shape of the velocity gradient. These changes are not unexpected: As the seeing worsens, a wider range of velocities from different parts of the galaxy is blurred together, and the emission from the core of the galaxy, which presumably has both the highest surface brightness and the lowest velocity, biases the velocity measurements out to larger and larger radii. Further, flux is lost as the seeing disk approaches and exceeds the size of the slit, the S/N declines as emission is spread over a larger area, and poorer seeing makes it more difficult to center the galaxy (or to be precise, the bright object we offset from) on the slit.

Q1700-BX691 (when observed under good conditions) shows the largest velocity shear we have detected so far; in order to assess the effects of the seeing on galaxies with a variety of velocity spreads, we have smoothed their spectra in order to simulate poorer seeing. This procedure was applied to all of the galaxies showing tilted lines in the May 2002 sample (with the exception of SSA22a-MD41, which was observed at higher spectral resolution with the ISAAC spectrograph on the VLT). We smoothed by first replacing the emission line in the two-dimensional spectra with the sky background (copied from a spatially adjacent portion of the slit, at the same wavelength) to make a sky frame, and then subtracting this sky frame from the original image to create an image of the emission line with the background removed. This profile was then smoothed with a Gaussian filter in the spatial direction; to simulate seeing of $0\rlap{.}''8$ with a spectrum in which the original seeing was $0\rlap{.}''5$, we used a filter with $0\rlap{.}''6$ FWHM ($0.5^2 + 0.6^2 = 0.8^2$). In order to account for slit losses due to the increase in seeing FWHM, we scaled the smoothed line by a factor of ~ 0.6, chosen empirically by measuring the decrease in line flux in the two observations of Q1700-BX691

---

[2]These data will be discussed in full elsewhere.

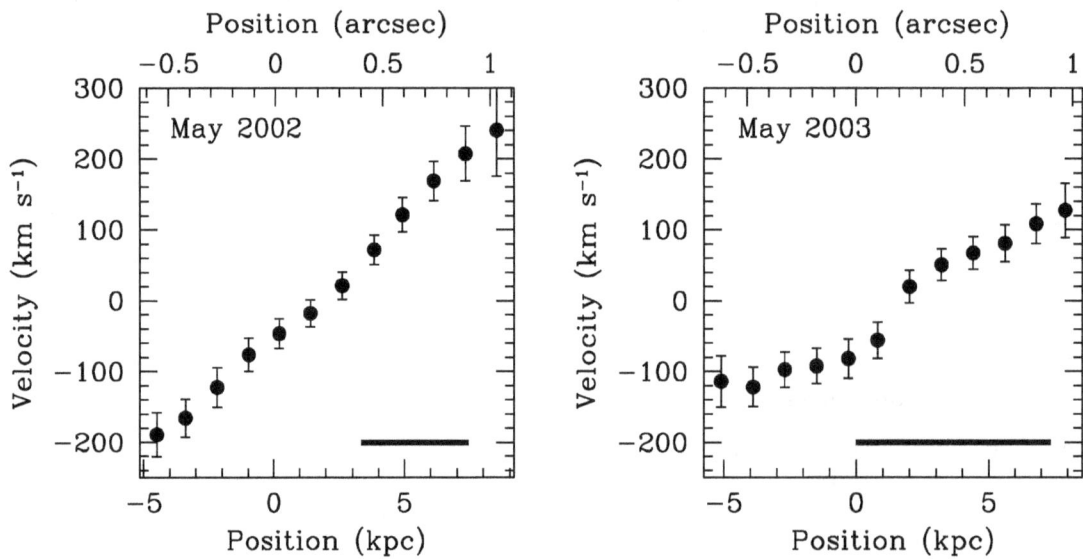

Figure 3.3 Velocity curves for Q1700-BX691, observed in May 2002 when the seeing was $\sim 0\!\!''5$ (left) and in May 2003 when the seeing was $\sim 0\!\!''9$ (right). The horizontal black lines represent the size of the seeing disk. The measured velocity amplitude decreases by a factor of ~ 2 when the galaxy is observed under poor conditions.

described above. The smoothed and scaled line was then added to the background frame to produce a final image. We then measured a velocity curve as usual on the resulting two-dimensional spectra. We find that the velocity spread is consistently reduced, but the detailed results are somewhat unpredictable. The effect of the smoothing depends not only on the initial tilt of the line, but also on the total flux and surface brightness. Lines with initial $v_c \lesssim 100$ km s^{-1} (Q1700-MD103, Q1623-BX511) no longer showed a tilt after smoothing, or had S/N too low to construct a velocity curve; those with larger v_c (West-BX600, Q1623-BX447) showed a decrease in the velocity spread of approximately a factor of 2, as we saw in the case of Q1700-BX691. The smoothed spectrum of Q1700-BX691 itself has $v_c \sim 125$ km s^{-1}, a good approximation to the May 2003 data, which have $v_c \sim 120$ km s^{-1}.

We use these results to place upper limits on the internal velocity spread of the ionized gas in the galaxies in the current sample: The 11 galaxies with no detected shear, and BX1332 with $v_c \sim 45$ km s^{-1}, are likely to have $v_{0.5} < 100$ km s^{-1}, where we use the term $v_{0.5}$ to indicate the velocity $(v_{max} - v_{min})/2$ that would be measured under seeing conditions of $\sim 0\!\!''5$. It is important to note here that the one remaining galaxy, BX1397, which shows the largest $v_c \sim 110$ km s^{-1}, was observed in July 2003 when the seeing was $0\!\!''5$ (as measured from a standard star observed immediately before the spectra were taken). Therefore all 13 of the galaxies have $v_{0.5} \leq 110$ km s^{-1}. In principle, of course, we would like to know the intrinsic velocity $v_{0.0}$ rather than $v_{0.5}$; this has been estimated with detailed modeling of rotation curves at lower redshift (Vogt et al. 1996, 1997). However, such a calculation requires a specific galaxy model such as an exponential disk, and knowledge of the galaxies' inclinations. Given the irregular morphologies of the galaxies in our sample, such assumptions are not justified. Even if we were to model an intrinsic velocity distribution, it is likely that the results would be highly degenerate; very different configurations of emitting material could produce the same tilted line when significantly degraded by the seeing. We therefore attempt to estimate only $v_{0.5}$.

The seeing clearly reduces our ability to resolve the velocity structure of a galaxy. However, it also mitigates the effects of slit misalignment, which under conditions of high spatial resolution can introduce spurious tilt in an emission line due to the correspondence between position of the emission within the slit and wavelength in the spectrum. For example, if the slit width is 200 km s^{-1} and a galaxy is tilted within the slit such that one end of it falls on one side of the slit and the other end on the other side of the slit, it will appear to have velocity shear of ± 100 km s^{-1} even if its actual velocity shear is zero. This problem will arise only when the FWHM of the galaxy light perpendicular to the slit is significantly smaller than the width of the slit, a situation that is unlikely to occur even under our best seeing conditions of $\sim 0\!\!''\!5$. Given the slight slit misalignments of the current observations, however, it is worthwhile to test the importance of this effect. We have done so by convolving the ACS image of each galaxy with a Gaussian filter to approximate the seeing of the NIRSPEC observation (under the assumption that the Hα emission traces the UV continuum), and measuring the centroid of the galaxy within the slit at each spatial position. We find that the galaxy light fills the slit in all cases, the maximum velocity introduced from the shift of the galaxy centroid across the slit is ± 15 km s^{-1}, and the mean induced velocity is ± 9 km s^{-1}; depending on the orientation of the galaxy, the velocity shear we observe could be either increased or decreased by these amounts. These numbers are less than our typical errors in velocity. We have also tested the effect of error in the position of the slit by measuring the smoothed galaxies' centroids within a simulated slit shifted by $0\!\!''\!2$ (our typical uncertainty in positioning the galaxy on the slit) in either direction perpendicular to the slit PA. Again we find that the effect is less than or comparable to our velocity errors, with a maximum introduced velocity of ± 21 km s^{-1} and a mean of ± 9 km s^{-1}. We therefore believe that only a small fraction of the velocity shear we observe in the tilted lines could be induced by misalignment of the slit.

A related concern involves the one-dimensional velocity dispersions. As noted above and discussed further in §3.5, the elongated galaxies for which we have placed slits on the extended axes have a lower average velocity dispersion than the sample as a whole. It is possible that by choosing random slit orientations we are artificially elevating the velocity dispersion by illuminating the slit unevenly as described above. We have tested this by smoothing the ACS images of known $z \sim 2$ galaxies to approximate $0\!\!''\!5$ seeing, placing simulated slits along them at random orientations, and measuring the shifts in the galaxies' centroids in the wavelength direction at each spatial position along the slit. For comparison with the velocity dispersion $\sigma = \mathrm{FWHM}/2.355$, we define $\sigma_{\mathrm{induced}} = \Delta v/2.355$, where Δv is the full velocity shift of the centroid. We find that the mean induced velocity dispersion $\langle \sigma_{\mathrm{induced}} \rangle = 8$ km s^{-1}, with a maximum of 21 km s^{-1}; these are nearly the same as the mean and maximum induced velocities found above, suggesting that even with large errors in position angle this effect does not introduce substantial velocity errors. The effect of $\sigma_{\mathrm{induced}}$ on the observed value of σ will depend on the source of the velocity dispersion: If the line widths are due to random motions, then the induced velocity dispersion will add to the true velocity dispersion in quadrature to produce the observed profile, and $\Delta\sigma = \sigma_{\mathrm{obs}} - \sigma_{\mathrm{true}}$ is only a few km s^{-1}. On the other hand, if the velocity dispersions are due to unresolved velocity shear, $\sigma_{\mathrm{induced}}$ could either increase or reduce the observed σ by up to ~ 20 km s^{-1}, an amount that is comparable to our typical error in σ. The effect on the average value of σ should be quite small, however, since these slit effects should increase or reduce σ with equal probability. The true situation is likely some combination of these two scenarios. In either case, however, the use of random position angles is unlikely to be a significant contaminant to our measurements of σ.

3.4.1.2 Velocity Dispersions

Although the seeing has a strong effect on the measurement of spatially resolved velocity shear, it has a relatively small effect on the one-dimensional velocity dispersion. This is because spatial resolution is not required to measure the velocity dispersion. We can see this by considering the limit in which the seeing disk exceeds the size of the galaxy. In this case we will be unable to distinguish emission from opposite sides of the galaxy (the line will no longer be tilted), but emission from both sides still contributes to the line width. We will then still measure the full velocity dispersion. The situation is slightly more complicated in practice, because of slit losses and decreased S/N: As the seeing worsens emission from the edges of the galaxy may fall outside the slit, decreasing the velocity dispersion somewhat, and S/N degraded by the seeing may limit the detection of faint wings in the line profile, especially if the galaxy's light is very centrally concentrated.

We have quantified these effects with the observations and simulations described above. For Q1700-BX691 we measure $\sigma = 170 \pm 18$ km s^{-1} with 0$\farcs$5 seeing, and $\sigma = 156 \pm 29$ km s^{-1} with 0$\farcs$9 seeing; the two values of σ agree within the errors. We find that this is generally the case with artificially smoothed data as well. The line widths are usually well-preserved after the smoothing, with $\gtrsim 80\%$ of the width retained; the weakest line (Q1623-BX511) shows a significant decrease, but also a substantial increase in the errors due to reduced S/N, such that the two values are still within $1\,\sigma$ of each other. The S/N of this degraded spectrum is lower than that of any object in the current sample. We therefore believe that our one-dimensional measurements are relatively uncompromised by the mediocre seeing conditions. The fact that the velocity spread of a tilted line is highly dependent on the seeing, while σ is not, makes it very difficult to estimate the former from measurements of the latter, as proposed by Weatherley & Warren (2003). Thus, their conclusion that Lyman break galaxies are preferentially low-mass star-bursting systems is premature.

3.4.2   Large-Scale Motions

Galactic-scale outflows, identified via the offsets between the redshifts of the UV interstellar absorption lines or Lyα and the nebular emission lines, are a common feature of galaxies at $z \sim 3$ (Pettini et al. 2001; Shapley et al. 2003). These outflows typically have speeds of ~ 300 km s^{-1} with respect to the systemic redshift of the galaxy, and are presumably powered by supernovae. A comparison of the redshifts given in Table 3.1 from the interstellar absorption lines, Lyα, and Hα shows that a similar pattern exists at $z \sim 2$. In Figure 3.4 we plot histograms of the velocity offsets of the interstellar absorption lines and Lyα with respect to Hα; we see that the mean offset of the absorption lines is $\langle \Delta v_{\mathrm{abs}} \rangle = -223 \pm 89$ km s^{-1}, and the mean offset of Lyα is $\langle \Delta v_{\mathrm{Ly}\alpha} \rangle = 470 \pm 116$ km s^{-1}. These velocities are consistent with those of all of the $z \sim 2$ galaxies for which we have performed this test (Steidel et al. 2004). Note that these velocities are measured from the centroids of the lines; we can estimate the terminal velocity of the outflows by adding half of the interstellar absorption line widths of ~ 650 km s^{-1} (Steidel et al. 2004) to the numbers given above.

Could these outflows be the cause of any of the tilted emission lines we have observed, or otherwise influence our kinematic measurements? The outflow velocities ($\langle \Delta v_{\mathrm{Ly}\alpha - \mathrm{abs}} \rangle \sim 700$ km s^{-1}) are several times larger than either the velocity spread of the tilted lines or the velocity dispersions, and the results of Shapley et al. (2003) show that nebular emission from HII regions indicates a galaxy's systemic redshift rather than the redshift of outflowing material: These authors see both stellar photospheric lines and nebular lines from HII regions in the composite spectrum of the $z \sim 3$ Lyman break galaxies, and find that their redshifts agree to within 50 km s^{-1}. Heckman, Armus, & Miley (1990) have shown that the emission line luminosity of the outflow in M82 is $\sim 10\%$ of the luminosity of the total galaxy. Calzetti et al. (2004) obtain a similar result in a study of the fraction of the ISM ionized by non-radiative processes in four local starburst galaxies, finding that 3–4% of the Hα luminosity arises from such shock-heated gas. If these relations hold in more distant galaxies, we would be unlikely to detect Hα from the outflows, especially given the strong redshift dependence of surface brightness. For these reasons we believe that the observed outflows are unlikely to influence our measurements of Hα.

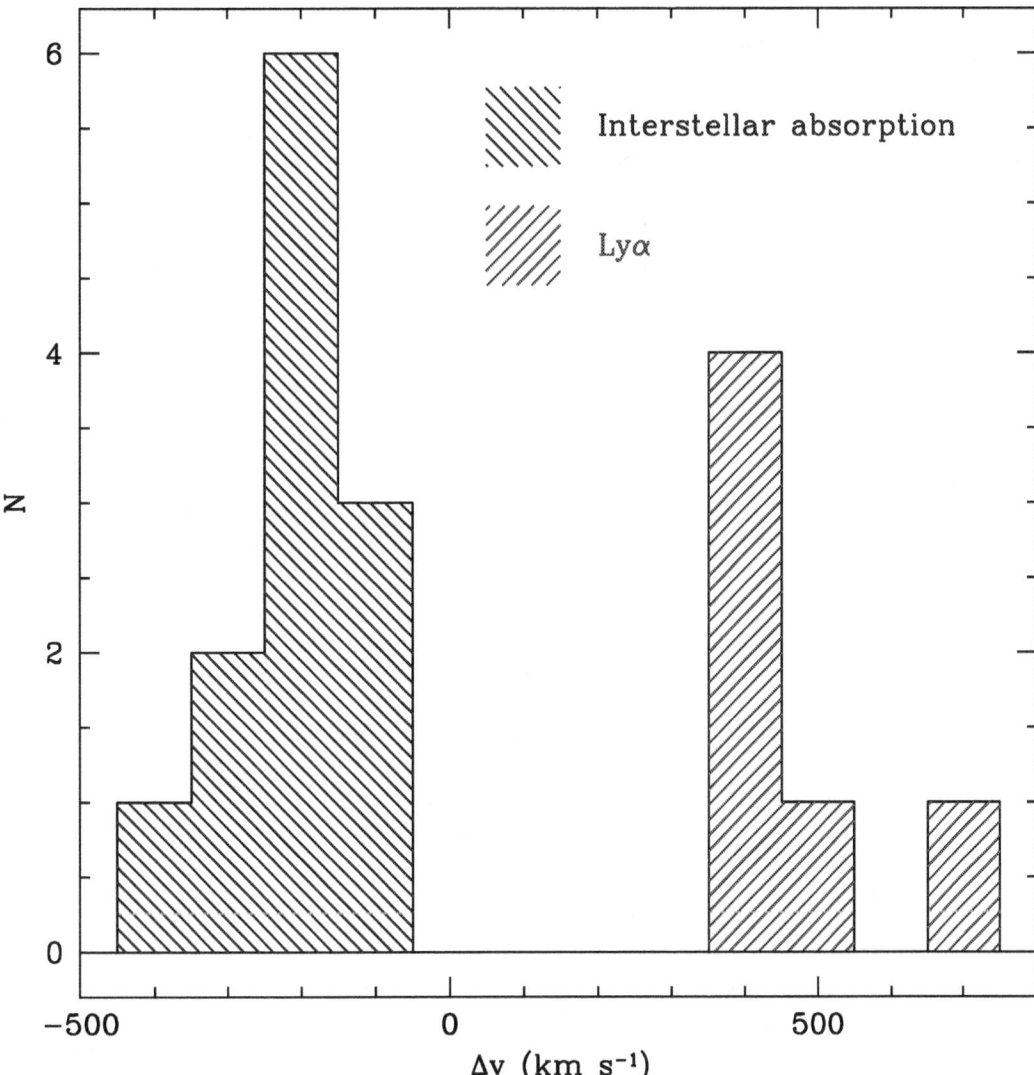

Figure 3.4 Velocity differences between the UV interstellar absorption lines and Hα (blue histogram at left) and between Lyα and Hα (red histogram at right). The offsets are consistent with galactic-scale superwinds with velocities of several hundred km s^{-1}.

3.5   Discussion

We have observed a lack of evidence for rotation in the form of tilted emission lines in elongated galaxies at $z \sim 2$, in spite of the fact that our spectroscopic slit was nearly aligned with the galaxies' major axes in 11 out of 13 cases. The seeing was mediocre during many of our observations, and we show with observations of the same object taken under different conditions and with artificially smoothed data that this can affect the observed velocity shear at a given radius by as much as a factor of 2. Accounting for the effects of the seeing, we believe that all 13 of the galaxies in the present sample have $v_{0.5} \leq 110$ km s^{-1}; 11 of them show no evidence of rotation at all. By $v_{0.5}$ we mean half of the peak-to-peak velocity amplitude measured in an exposure of ~ 1 hour under good seeing conditions ($\sim 0''.5$); unlike the terminal velocities of rotation curves of local galaxies, this is not a fundamental quantity, and it likely underestimates the true velocity. This observed lack of rotation is perhaps surprising; rotational velocities of a few hundred km s^{-1} are predicted for models of star-forming galaxies at $z \sim 2$–3 (Mo, Mao, & White 1998, 1999), and we know that galaxies with $v_c \sim 200$ km s^{-1} are present among the still relatively small sample with rest-frame optical emission line spectra at these redshifts (Lemoine-Busserolle et al. 2003; Erb et al. 2003).

How, then, are we to interpret the lack of observed rotation in galaxies with slits aligned with their apparent major axes? The kinematic measurements indicate that most of the galaxies are not disks with irregular morphologies due to knotty star formation. In fact we find that the alignment of the slit with a galaxy's major axis has little effect on whether or not a tilted emission line is observed. Of our total sample of 29 galaxies (the 13 presented here and the 16 in Erb et al. 2003), 12 had slits aligned with their major axes and 17 did not. We see tilted emission lines in 25% of the aligned galaxies and 29% of the unaligned galaxies; it seems that we are slightly more likely to observe a tilted line by choosing a random position angle, although this is hardly a robust result given the small numbers involved and the effect of the seeing on the detectability of velocity shear.

Given these results, we should consider the possibility that the tilted emission lines, when observed, may not always be due to rotation. Two or more merging clumps of material with relative velocities of ~ 100 km s^{-1}, blurred by the seeing, could produce an extended, tilted emission line. Such a scenario fits naturally within the context of hierarchical structure formation, and has been used to explain the profiles of damped Lyα absorption systems at high redshift (Haehnelt, Steinmetz, & Rauch 1998). In this model the measured velocity would depend on the orientation of the merging clumps, and projection effects would be difficult to quantify. This is not meant to be an exclusive explanation for the origin of the tilted lines; rotation could certainly still be a factor as well, especially in the case of objects such as Q1700-BX691 that show a regular pattern of high velocity shear over a large spatial area.

The uncertain causes and unreliable velocities of the tilted lines make inferences from this type of kinematic measurement difficult, and it therefore makes sense to focus our attention on what can be learned from the one-dimensional measurements, which are relatively unaffected by the seeing. In our simulations at least 80% of the velocity width was retained after smoothing in all except one case; after the smoothing, the exception had S/N lower than that of any object in the current sample. We have therefore compared the velocity dispersions of the sample with their morphologies; we see some evidence that the more elongated galaxies may have smaller velocity dispersions than the sample as a whole. Considering only those galaxies that we selected because they appeared elongated (i.e., neglecting BX305, BX1085, BX1086, and BX1368), we find a mean velocity dispersion of 80 km s^{-1}, with an error in the mean of 9 km s^{-1}. We compare this with the mean velocity dispersion of the remaining 54 galaxies at $z \sim 2$ for which we have Hα measurements (these data will be discussed in full elsewhere), $\langle \sigma \rangle = 118 \pm 7$ km s^{-1}; the difference in the means is more than 3 σ. Using a K-S test, we find that the probability that the two samples are drawn from the same population is 0.04. To investigate this further we plot in Figure 3.5 the aspect ratio $a = d_{min}/d_{maj}$ against the velocity dispersion σ. There is mild evidence for a correlation, in the sense that the more symmetric galaxies also have larger velocity dispersions; with a Spearman rank-order correlation test we find that the probability of observing these data if there is no correlation is 0.13.

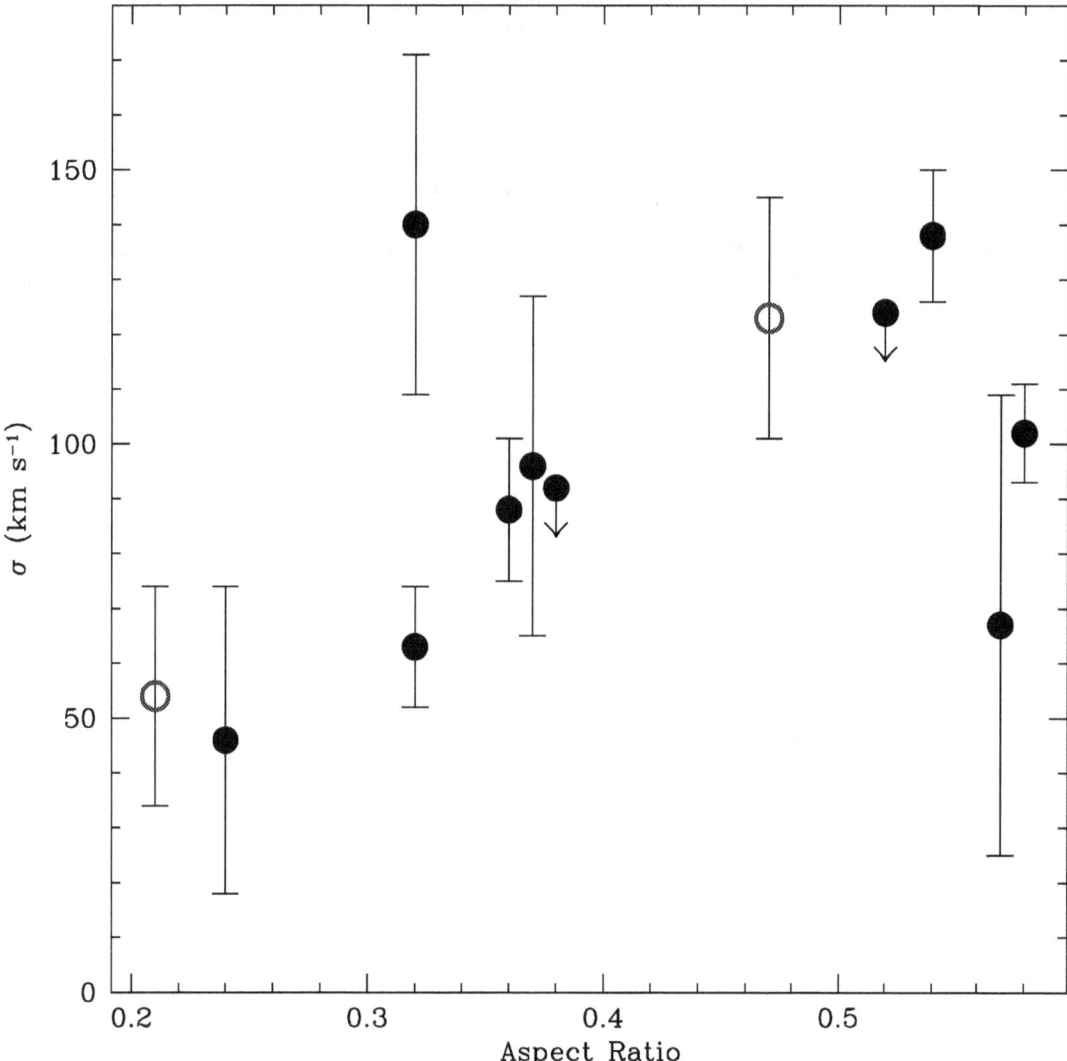

Figure 3.5 The velocity dispersion σ plotted against the aspect ratio d_{min}/d_{maj}. The two open red symbols are those galaxies with tilted emission lines. We see mild evidence that the more symmetric galaxies have larger velocity dispersions.

These nine galaxies have $\langle a \rangle = 0.38$, with an error in the mean of 0.04. We compare this with the 90 other galaxies we have identified at $2.0 < z < 2.7$ in the GOODS-N field. These have $\langle a \rangle = 0.51$, with an error in the mean of 0.02; the difference in the means is 3σ, showing that this subsample is significantly more elongated than one made up of galaxies chosen at random. We have also calculated the mean $E(B-V)$ (as determined from the $G - \mathcal{R}$ colors, assuming a star formation age of $> 10^7$ yrs and a Salpeter IMF) of the elongated sample, $\langle E(B - V) \rangle = 0.105 \pm 0.017$ (we again quote the mean and the error in the mean). As above we compare this with the remaining galaxies with $2.0 < z < 2.7$ in the GOODS-N field, which have $\langle E(B - V) \rangle = 0.149 \pm 0.009$, and find that the elongated galaxies have a lower $\langle E(B - V) \rangle$ by 2σ. These nine galaxies are also somewhat brighter than the sample as a whole, with $\langle \mathcal{R} \rangle = 23.87 \pm 0.15$ as compared to $\langle \mathcal{R} \rangle = 24.32 \pm 0.06$;[3] we see no correlation between $\mathcal{R}$ magnitude and $E(B - V)$, however. Both the decreased reddening and the lower velocity dispersions are the opposite of what one would expect if these galaxies were highly inclined disks: Increasing the inclination of a disk also makes it appear redder, and the alignment of the slit and the major axis should maximize the observed velocity.

We have seen that elongated galaxies sometimes, but not always, show evidence for velocity shear along their major axes, and that clumpy galaxies that appear merger-like sometimes, but not always, show evidence of shear as well. We also see shear in galaxies of unknown morphology with the slit misaligned by an unknown amount. The situation is clearly complicated, and it is unlikely that a single model will explain all of these results; some combination of rotation and merging seems to be the most likely answer. It will be challenging to distinguish between these possibilities with the type of measurements presented here, however. Even under ideal conditions, the blurring of the seeing makes it difficult to discriminate between rotation and merging. Evidence for rotation could be found in deeper observations if the velocity curves are seen to flatten like the rotation curves of local galaxies, and more detailed morphological analysis may help pick out galaxies that are likely to be mergers. The best way to distinguish between the proposed scenarios, however, would be to obtain spectroscopic observations of an entire galaxy at high spatial resolution, and we look forward to observations of objects such as these with the new generation of near-IR integral field spectrographs with adaptive optics.

We thank the anonymous referee for useful comments. We would also like to thank the staff at the Keck Observatory for their competent assistance with our observations. CCS, DKE, and AES have been supported by grants AST00-70773 and AST03-07263 from the U.S. National Science Foundation and by the David and Lucile Packard Foundation. AES acknowledges support from the Miller Institute for Basic Research in Science. Finally, we wish to extend special thanks to those of Hawaiian ancestry on whose sacred mountain we are privileged to be guests. Without their generous hospitality, the observations presented herein would not have been possible.

---

[3]Because our only photometry is in the rest-frame UV, we defer the calculation of absolute magnitudes until we have obtained rest-frame optical photometry.

Table 3.1. Galaxies Observed

Galaxy	R.A. (J2000)	Dec. (J2000)	Exposure time (s)	z_{abs} [a]	$z_{Ly\alpha}$ [b]	$z_{H\alpha}$ [c]	$\mathcal{R}$	$G - \mathcal{R}$	$F_{H\alpha}$ [d]	Selection [e]
BX305	12:36:37.131	62:16:28.358	900 × 4	2.4825	–	2.4839	24.28	0.79	4.2	R
BX1055	12:35:59.594	62:13:07.504	900 × 2	2.4865	2.4959	2.4901	24.09	0.24	2.7	E
BX1084	12:36:13.568	62:12:21.485	900 × 5	2.4392	–	2.4403	23.24	0.26	7.3	E
BX1085	12:36:13.331	62:12:16.310	900 × 5	2.2381	–	2.2407	24.50	0.33	1.1	T
BX1086	12:36:13.415	62:12:18.841	900 × 5	–	–	2.4435	24.64	0.41	1.8	T
BX1277	12:37:18.595	62:09:55.536	900 × 3	2.2686	–	2.2713	23.87	0.14	5.3	E
BX1311	12:36:30.540	62:16:26.116	900 × 4	2.4804	2.4890	2.4843	23.29	0.21	8.0	E
BX1322	12:37:06.538	62:12:24.938	900 × 6	2.4401	2.4491	2.4443	23.72	0.31	2.0	E
BX1332	12:37:17.134	62:11:39.946	900 × 3	2.2113	–	2.2136	23.64	0.32	4.4	E
BX1368	12:36:48.241	62:15:56.237	900 × 4	2.4380	2.4455	2.4407	23.79	0.30	8.8	R
BX1376	12:36:52.960	62:15:45.545	900 × 4	2.4266	2.4338	2.4294	24.48	0.01	2.2	E
BX1397	12:37:04.115	62:15:09.837	900 × 3	2.1322	–	2.1332	24.12	0.14	5.2	E
BX1479	12:37:15.417	62:16:03.876	900 × 5	2.3726	2.3823	2.3745	24.39	0.16	2.5	E

[a]Vacuum heliocentric redshift from rest-frame UV interstellar absorption lines.

[b]Vacuum heliocentric redshift of Lyα emission line, when present.

[c]Vacuum heliocentric redshift of Hα emission line.

[d]Flux of Hα emission line, in units of 10^{-17} ergs s^{-1} cm^{-2}. This should be considered a lower limit because conditions were non-photometric.

[e]Reason for selection. E=elongated, T="triplet," additional galaxies on slit with BX1084, R=favorable redshift. See §2 for details.

Table 3.2.   Morphologies and Kinematics

Galaxy	$z_{H\alpha}$[a]	PA[b]	Slit PA[c]	$r_{1/2}$[d] (arcsec)	$r_{1/2}$[d] (kpc)	d_{maj}[e] (arcsec)	d_{maj}[e] (kpc)	a[f]	$d_{H\alpha}$[g] (arcsec)	$d_{H\alpha}$[g] (kpc)	σ[h] (km s^{-1})	v_c[i] (km s^{-1})	M_{vir}[j] (10^{10} M$_\odot$)
BX305	2.4839	109.2	133.4	0.28	2.3	0.25	2.0	0.32	–	–	140 ± 31	–	4.1
BX1055	2.4901	107.0	120.0	0.14	1.2	0.09	0.7	0.52	0.48	3.8	< 124	–	1.2
BX1084	2.4403	−0.6	18.7	0.24	1.9	0.16	1.3	0.58	0.59	4.8	102 ± 9	–	1.5
BX1085	2.2407	93.3	18.7	0.29	2.4	0.25	2.0	0.39	0.27	2.2	–	–	–
BX1086	2.4435	−19.1	18.7	0.13	1.1	0.08	0.6	0.57	–	–	67 ± 42	–	0.25
BX1277	2.2713	49.8	58.1	0.26	2.1	0.22	1.8	0.32	0.62	5.1	63 ± 11	–	0.83
BX1311	2.4843	47.2	55.3	0.26	2.1	0.33	2.7	0.36	1.14	9.2	88 ± 13	–	2.3
BX1322	2.4443	47.7	57.1	0.24	1.9	0.19	1.5	0.38	0.57	4.6	< 92	–	1.5
BX1332	2.2136	−29.9	−22.0	0.32	2.6	0.37	3.1	0.21	1.47	12.1	54 ± 20	45	1.0
BX1368	2.4407	78.8	90.3	0.15	1.2	0.09	0.7	0.54	0.90	7.3	138 ± 12	–	1.5
BX1376	2.4294	59.7	59.6	0.24	2.0	0.20	1.6	0.37	0.22	1.8	96 ± 31	–	1.8

Table 3.2—Continued

Galaxy	$z_{H\alpha}$ [a]	PA [b]	Slit PA [c]	$r_{1/2}$ [d] (arcsec)	$r_{1/2}$ [d] (kpc)	d_{maj} [e] (arcsec)	d_{maj} [e] (kpc)	a [f]	$d_{H\alpha}$ [g] (arcsec)	$d_{H\alpha}$ [g] (kpc)	σ [h] (km s^{-1})	v_c [i] (km s^{-1})	M_{vir} [j] (10^{10} M$_\odot$)
BX1397	2.1332	−37.4	−31.0	0.30	2.5	0.20	1.6	0.47	1.02	8.4	123 ± 22	110	2.8
BX1479	2.3745	−0.5	9.7	0.22	1.8	0.23	1.9	0.24	0.51	4.2	46 ± 28	–	0.45

[a]Vacuum heliocentric redshift of Hα emission line.

[b]Position angle of galaxy in degrees.

[c]Position angle of slit in degrees.

[d]Effective half-light radius; see §3.3.

[e]RMS dispersion of light about centroid along major axis; see §3.3.

[f]Aspect ratio of RMS dispersions along minor and major axes; see §3.3.

[g]Approximate spatial extent of Hα emission line.

[h]One-dimensional velocity dispersion from Hα emission line.

[i]$(v_{max} - v_{min})/2$ for those galaxies that show tilted emission lines.

[j]Virial masses calculated from σ and d_{maj}; see §3.3 and §3.4.

Chapter 4

The Rest-Frame Optical Properties of Star-Forming Galaxies at $z \sim 2$:
Kinematics and Star Formation from Hα Spectroscopy and Ultraviolet to Infrared Imaging[*]

Dawn K. Erb,[a] Charles C. Steidel,[a] Alice E. Shapley,[b] Max Pettini,[c] Naveen A. Reddy,[a] Kurt L. Adelberger[d]

[a]California Institute of Technology, MS 105–24, Pasadena, CA 91125

[b]Department of Astronomy, 601 Campbell Hall, University of California at Berkeley, Berkeley, CA 94720

[c]Institute of Astronomy, Madingley Road, Cambridge CB3 0HA, UK

[d]Carnegie Observatories, 813 Santa Barbara Street, Pasadena, CA 91101

Abstract

We present analysis of the Hα spectra of 114 rest-frame UV-selected star-forming galaxies at $z \sim 2$. Using rest-frame UV to optical or IR imaging we model the stellar populations of $\sim 80\%$ of the galaxies in the sample, in order to examine the Hα properties as a function of stellar mass and age. The Hα line widths give a mean velocity dispersion $\sigma = 112$ km s^{-1}, and a mean dynamical mass $M_{\rm dyn} = 4.3 \times 10^{10}$ M$_{\odot}$, after excluding AGN. The average dynamical mass is ~ 2 times larger than the average stellar mass, and the two are correlated at the $3\,\sigma$ level and agree to within a factor of several for most objects, consistent with observational and systematic uncertainties. However, $\sim 15\%$ of the sample has $M_{\rm dyn} \gg M_{\star}$. These objects are best fit by young stellar populations and tend to have high Hα equivalent widths, suggesting that they are young starbursts with large gas masses. Rest-frame optical luminosity and velocity dispersion are correlated with 4σ significance, though the correlation has far more scatter than the local equivalent, and more than can be accounted for by the uncertainties. Fourteen of the 114 galaxies in the sample have spatially resolved and tilted Hα emission lines indicative of velocity shear. It is not yet clear whether the shear indicates merging or rotation, but if the galaxies are rotating disks and follow relations between velocity dispersion and circular velocity similar to those seen in local galaxies, we underestimate the circular

---

[*]Based on data obtained at the W.M. Keck Observatory, which is operated as a scientific partnership among the California Institute of Technology, the University of California, and NASA, and was made possible by the generous financial support of the W.M. Keck Foundation.

velocities by an average factor of ~ 2 and the sample has $\langle V_c \rangle \sim 190$ km s^{-1}. The presence of galactic-scale outflows is confirmed through the offsets between the nebular, interstellar absorption, and Lyα emission lines, but there is no significant correlation between the outflow speeds and stellar or dynamical mass, line width, or star formation rate. There may be a connection between decreased blueshift of the interstellar absorption lines and the presence of velocity shear. The average star formation rate (SFR) from extinction-corrected Hα luminosity is ~ 30 M$_\odot$ yr^{-1}, in good agreement with SFRs from both the rest-frame UV continuum and X-rays. SFR increases with stellar mass and at brighter K magnitudes; galaxies with $K_s < 20$ have $\langle \mathrm{SFR}_{\mathrm{H}\alpha} \rangle \sim 70$ M$_\odot$ yr^{-1}. Using the local empirical correlation between star formation rate per unit area and gas surface density, we estimate the mass of the gas associated with star formation, and find a strong increase in gas fraction with decreasing stellar mass. The masses of gas and stars combined are considerably better correlated with the dynamical masses than are the stellar masses alone. The combination of kinematic measurements, estimates of gas masses, and SED modeling suggests that the factor of ~ 500 range in stellar mass in our sample is due as much to the evolution of the stellar population and the conversion of gas into stars as to intrinsic differences in the total masses of the galaxies.

4.1   Introduction

The rest-frame optical spectra of galaxies are among the most powerful tools for understanding their modes of star formation, their kinematics, and their chemical abundances. Such techniques have been used to study local galaxies almost since it became known that there were other galaxies, but for redshifts beyond $z \sim 1.5$, the strong nebular emission lines of star-forming galaxies shift into the infrared, making such studies much more difficult. Historically, this meant that galaxies in the $1.5 \lesssim z \lesssim 2.5$ range were difficult to identify. In the past several years, however, criteria have been developed that efficiently select such objects based on their rest-frame UV (Steidel et al. 2004) or optical (Franx et al. 2003; Daddi et al. 2004) colors. These criteria, in combination with improvements in both blue-sensitive and near-IR spectrographs, have led to vast increases in both the number of galaxies known at these redshifts and in quantitative studies of their properties.

Because of the not insignificant observing time required per object, even on the largest telescopes, and the lack of multi-object near-IR spectrographs, samples of rest-frame optical spectra at high redshift have so far been small, limited to a few to ~ 20 objects, though the number of such studies has rapidly been increasing (Pettini et al. 2001; Lemoine-Busserolle et al. 2003; Moorwood et al. 2003; Erb et al. 2003, 2004; van Dokkum et al. 2004; Shapley et al. 2004; Swinbank et al. 2004). Nevertheless, these studies have provided significant insight into the star formation rates, kinematic properties, and metallicities of galaxies at high redshift. For example, the widths of nebular emission lines provide direct estimates of mass that depend little on stellar population properties, though these are complicated by the unknown mass distribution of galaxies at high redshift, by the difficulty of obtaining accurate measurements of sizes, and by uncertainties in the extent to which the nebular lines trace the distribution of mass. Results so far suggest that typical UV-selected galaxies have dynamical masses of a few 10^{10} M$_\odot$ (Pettini et al. 2001; Erb et al. 2003, 2004; Shapley et al. 2004), while galaxies selected by their red $J - K$ colors have dynamical masses $\sim 10^{11}$ M$_\odot$ (van Dokkum et al. 2004; this conclusion depends on the apparently large sizes of the red galaxies as well as on their large line widths, and we note that there are similar objects in the UV-selected sample).

Accurate systemic redshifts from the nebular emission lines have also been instrumental in identifying the galactic-scale winds that are omnipresent in star-forming galaxies at high redshift (Pettini et al. 2001; Adelberger et al. 2003; Adelberger et al. 2005a). The presence of such winds is inferred through offsets of several hundred km s^{-1} between the nebular emission lines and the interstellar absorption lines and Lyα emission seen in the rest-frame UV spectra, as well as through their effect on the surrounding IGM. Studies have so far been limited, but the properties of the outflows have not been seen to significantly depend on the properties of the galaxies themselves (Adelberger et al. 2005a), though such correlations have been seen at low redshift (Martin 2005; Rupke et al. 2005), where a much wider range of galaxies' properties is accessible.

Observations of Hα at $z \sim 2$ indicate uniformly high star formation rates (compared to more quiescent local galaxies) of at least a few $M_\odot$ yr^{-1} and often much higher; this is true by definition, of course, since significant star formation is required to make Hα detectable in a (non-AGN) galaxy at high redshift. Galaxies with bright K magnitudes have been found to have especially high SFRs of tens to hundreds of $M_\odot$ yr^{-1} (Shapley et al. 2004; van Dokkum et al. 2004), consistent with a trend of increasing SFR at brighter K magnitudes for all samples of galaxies at $z \sim 2$ found by Reddy et al. (2005), using stacked X-ray images. The use of Hα as a star formation indicator is particularly valuable at high redshift because it is widely used in local galaxies, which thus provide an important sample for both comparison and calibration (e.g., Kennicutt 1998a; Brinchmann et al. 2004). In this sense Hα is considerably more useful than the rest-frame UV continuum, which is easily obtained from optical images of high redshift galaxies but more difficult to observe locally. Nearly all galaxies with Hα spectra at high redshift have rest-frame UV images as well, however, so the UV continuum provides a direct comparison for the Hα-derived SFRs.

Further insight into the properties of high redshift galaxies has been gained with multi-wavelength imaging extending to the rest-frame optical or near-IR, which allows modeling of the spectral energy distribution and thus provides constraints on the stellar populations (e.g., Papovich et al. 2001; Shapley et al. 2001; Förster Schreiber et al. 2004; Shapley et al. 2005b). Such models suffer from degeneracies between age and extinction but can usually determine stellar masses to a factor of a few or better, provided, of course, that the star formation episode being modeled represents the total stellar population of the galaxy. As spectroscopy is not required for such models (except to confirm redshifts), large samples have been relatively easy to obtain, and indeed large samples are required both to begin to overcome the uncertainties in the modeling and obtain statistically meaningful results, and to assess the diversity of stellar populations among the galaxies in question. The combination of near-IR spectra and model SEDs to assess the properties of high redshift galaxies has so far focused primarily on objects with large stellar masses, which have high metallicities in addition to above-average star formation rates and line widths (Shapley et al. 2004; van Dokkum et al. 2004).

In this paper we present an enlarged sample of Hα spectra of 114 $z \sim 2$ galaxies selected by their rest-frame UV colors. We combine the Hα results with model SEDs derived from optical to near- or mid-IR imaging, in order to assess kinematic and star formation properties as a function of stellar mass. The paper is organized as follows: We describe the selection of our sample, the observations, and our data reduction procedures in §4.2. In §4.3 we outline the modeling procedure by which we determine stellar masses and other stellar population parameters. Section 4.4 is devoted to the galaxies' kinematics. In §4.4.1 we derive dynamical masses from the Hα line widths, compare them with stellar masses, and assess the relationship between velocity dispersion and rest-frame optical luminosity. We compare the distributions of velocity dispersions of galaxies at $z \sim 2$ and $z \sim 3$ in §4.4.1.3. We examine galaxies with spatially resolved and tilted emission lines in §4.4.2, and discuss galactic outflows in §4.4.3. In §4.5 we discuss star formation rates derived from the luminosities of Hα and the UV continuum. Also in this section (§4.5.2) we estimate gas masses from the local correlation between star formation rate and gas density per unit area, and compare stellar, gas, and dynamical masses. We summarize our conclusions and discuss our results in §4.6. Separately, we use the same sample of Hα spectra to construct composites according to stellar mass to show that there is a strong correlation between increasing oxygen abundance as measured by the [N II]/Hα ratio and increasing stellar mass (Erb et al. 2005a).

A cosmology with $H_0 = 70$ km s^{-1} Mpc^{-1}, $\Omega_m = 0.3$, and $\Omega_\Lambda = 0.7$ is assumed throughout. In such a cosmology, $1''$ at $z = 2.24$ (the mean redshift of the current sample) corresponds to 8.2 kpc, and at this redshift the universe is 2.9 Gyr old, or 21% of its present age. For calculations of stellar masses and star formation rates we use a Chabrier (2003) initial mass function (IMF), which results in stellar masses and SFRs 1.8 times smaller than would be obtained with a Salpeter (1955) IMF.

4.2   Sample Selection, Observations, and Data Reduction

The galaxies discussed herein are drawn from the rest-frame UV-selected $z \sim 2$ spectroscopic sample described by Steidel et al. (2004). The candidate galaxies are selected by their $U_nG\mathcal{R}$ colors (from deep optical images discussed by Steidel et al. 2004), with redshifts then confirmed in the rest-frame UV using the LRIS-B spectrograph on the Keck I telescope on Mauna Kea. Our total sample of 114 Hα spectra consists of 75 new observations and 39 that have been published previously (Erb et al. 2003, 2004; Shapley et al. 2004). This set of galaxies is not necessarily representative of the UV-selected sample as a whole, because objects were chosen for near-IR spectroscopy for a wide variety of reasons. Approximately 20% of the galaxies were selected because they lie near the line of sight to a QSO (for studies of correlations between galaxies and metal systems seen in absorption in the QSO spectra; see Adelberger et al. 2005a). Others ($\sim$ 15%) were chosen based on their morphologies (elongated in most cases, with a few more compact objects for comparison; most of these have been previously discussed by Erb et al. 2004) in the HST ACS images of the GOODS-N field. Some objects were selected because of their red or bright near-IR colors or magnitudes, or occasionally because their photometry suggested an unusual SED (13 objects selected because they have $K < 20$ are discussed by Shapley et al. 2004). We observed some galaxies because of their excellent deep rest-frame UV spectra, or occasionally simply because they were bright in the rest-frame UV. Several galaxies were observed to confirm their classification as AGN. A few objects were observed because they were close pairs with favorable redshifts with respect to the night sky lines. The remaining objects are galaxies within $\sim 20''$ of our primary targets, which were placed along the slit and observed simultaneously with the primary targets. A significant number of galaxies meet more than one of these criteria. The sightline and secondary pair objects are therefore generally representative of the UV-selected sample as a whole, while the addition of the IR red or bright objects means that the total Hα sample is somewhat biased toward more massive galaxies (see §4.3). The AGN fraction (5/114) is similar to the fraction found for our full spectroscopic sample in the GOODS-N field by Reddy et al. (2005), using direct detections in the 2-Ms *Chandra* Deep Field North images.

The Hα detection rate for galaxies with previously known redshifts is $\sim$ 80%. Not including objects for which there was a known problem with the NIRSPEC rotator or incorrect astrometry in one field, we observed 23 such galaxies without detecting them. For objects that meet our photometric criteria but do not have previously determined redshifts the detection rate is much lower. We have observed 28 such galaxies and detected five, for a rate of 18% (we require a detection of greater significance when the redshift is not previously known). Some insight into the reasons for the non-detections can be gained by looking at the objects that were undetected but placed on the slit with another object that was detected; this gives us confidence that the non-detections were not caused by incorrect pointing or tracking problems. There are 11 such objects with previously known redshifts. Two of these 11 are undetected in K, and six of the nine remaining have $K > 21.5$. Comparing with our NIRSPEC sample, we see that 15% of the galaxies with Hα detections have $K > 21.5$ (the average K magnitude of our NIRSPEC sample is 20.7), while 73% of the non-detections meet this condition. We are clearly less likely to detect Hα emission for objects that are faint in K. This is not surprising, given the correlation between K magnitude and star formation rate recently found for the $z \sim 2$ sample by Reddy et al. (2005). We confirm and discuss this correlation in §4.5. Not all of our non-detections can be ascribed to intrinsically low flux levels, however, as shown by the fact that three initially non-detected objects were detected on a second observation (out of six that were reobserved). Other reasons for non-detections may include telescope pointing or tracking error, poor observing conditions, or offsets between the position of Hα emission and the continuum flux used for astrometry.

For the purposes of comparisons with other surveys, we note that 10 of the 87 galaxies for which we have JK_s photometry have $J - K_s > 2.3$ (the selection criterion for the FIRES survey, Franx et al. 2003); this is similar to the $\sim 12\%$ of UV-selected galaxies that meet this criterion (Reddy et al. 2005). Eighteen of the 93 galaxies for which we have K magnitudes have $K_s < 20$, the selection criterion for the K20 survey (Cimatti et al. 2002); this is a higher fraction than is found in the full UV-selected sample ($\sim 10\%$), because we intentionally targeted many K-bright galaxies (Shapley et al. 2004). Five of the 10 galaxies with $J - K_s > 2.3$ also have $K_s < 20$.

4.2.1 Near-IR Spectra

All of the Hα spectra (with the exceptions of CDFb-BN88, Q0201-B13, and SSA22a-MD41, which were observed with the ISAAC spectrograph on the VLT and previously discussed by Erb et al. 2003) were obtained using the near-IR spectrograph NIRSPEC (McLean et al. 1998) on the Keck II telescope. Observing runs were in May 2002; May, July and September 2003; and June and September 2004. The conditions were generally photometric with good (~ 0.5–$0.6''$) seeing, with the exceptions of the May 2003 and June 2004 runs, which suffered from occasional clouds and seeing up to $\sim 1''$. Hα falls in the K-band for the redshift range $2.0 \lesssim z \lesssim 2.6$, which includes the vast majority of the current sample; 6 of the 114 objects have $1.4 \lesssim z \lesssim 1.8$, and were observed in the H-band. We use the $0.76'' \times 42''$ slit, which, in NIRSPEC's low resolution mode, provides a resolution of ~ 15 Å (~ 200 km s^{-1}; $R \simeq 1400$) in the K-band.

Our observing procedure is described by Erb et al. (2003). For a typical object, we use four 15 minute integrations, although the number of exposures varies from 2 to 6 for total integrations of 0.5 to 1.5 hours. We perform blind offsets from a nearby bright star, returning to the offset star between each integration on the science target to recenter and dither along the slit. In most cases we attempt to observe two galaxies with separation $< 20''$ simultaneously by placing them both on the slit; thus the position angle is set by the positions of the two galaxies. The objects observed are listed in Table 4.1.

The spectra are reduced using the standard procedures we have previously described (Erb et al. 2003). The most difficult step in such reductions is the absolute flux calibration, which is subject to significant uncertainties, primarily due to slit losses from the seeing and imperfect centering of the object on the slit. Because the exposures of the standard stars we use as references are not usually taken immediately before or after the science targets (primarily because the NIRSPEC detector suffers from charge persistence after observations of bright objects), our calibration may also be affected by differences in seeing and weather conditions between the science and calibration observations. Several methods have been used to assess the accuracy of our flux calibration. Using a narrow-band image of the Q1700 field (centered on Hα at $z = 2.3$, for observations of the proto-cluster described by Steidel et al. 2005), we have measured narrowband Hα fluxes for six of the objects in our sample, and find that the NIRSPEC Hα fluxes are $\sim 50\%$ lower. For those (relatively few) galaxies for which we detect significant continuum in the NIRSPEC spectra, we can compare the average flux density in the K band with our broadband magnitudes. These tests also indicate that our NIRSPEC fluxes are low by a factor of ~ 2 or more. We have also assessed the effects of losses from the slit and the aperture used to extract the spectra by constructing a composite two-dimensional spectrum of all the objects in the sample and comparing its spatial profile to the widths of the slit and our aperture. This test suggests losses of $\sim 40\%$, although this figure represents a lower limit because our procedure of dithering the object along the slit and subtracting adjacent images results in the occasional loss of flux from extended wings. Motivated by these tests, we have when noted applied a factor of 2 aperture correction for the determination of star formation rates and Hα equivalent widths. Although the correction is imprecise, especially for individual objects, we believe that applying the correction results in a closer approximation to the true average flux of the sample than leaving the fluxes uncorrected.

4.2.2   Near-IR Imaging

J-band and K_s-band images were obtained with the Wide-field IR Camera (WIRC; Wilson et al. 2003) on the 5 m Palomar Hale telescope, which uses a Rockwell HgCdTe Hawaii-2 2k × 2k array, with a field of view of $8.5' \times 8.5'$ and spatial sampling of 0.249 arcseconds per pixel. Observations were conducted in June and October 2003 and April, May, and August 2004. Some images were also taken in June 2004, courtesy of A. Blain and J. Bird. We used 120 second integrations (four 30 second coadds in K, one 120 second coadd in J), typically in a randomized $\sim 8''$ dither pattern of 27 exposures. Total exposure times and 3σ image depths in each field are given in Table 4.2.

Only those images with seeing as good as or better than our optical images (typically $\sim 1.2''$; $0.85''$ in the Q1700 field) were incorporated into the final mosaics. The images from each dither sequence were reduced, registered, and stacked using IDL scripts customized for WIRC data by K. Bundy (private communication). The resulting images were then combined and registered to our optical images using IRAF tasks. Flux calibration was done with reference to the 15–30 2MASS stars in each field. We have also used a 45 minute image of Q2346-BX404 and Q2346-BX405 from the near-IR camera NIRC (Matthews & Soifer 1994) on the Keck I telescope taken on 1 July 2004 (UT), courtesy of D. Kaplan and S. Kulkarni; this image was reduced similarly, and flux-calibrated by matching the magnitudes of objects in the field to their magnitudes in the calibrated Q2346 WIRC image.

Photometry was performed as described by Shapley et al. (2005b), with the difference that we detected the objects in both the $\mathcal{R}$ and K images, and after applying both sets of isophotes to the K and J images used whichever was the more significant of the two detections. Note that this change in method, as well as a slight change in the photometric zeropoints from new flux calibrations, means that some magnitudes given here may be slightly different than those previously published (Shapley et al. 2004, 2005b). The K-band images have been trimmed somewhat more than previous versions for cleaner detections in K, with the result that some objects with previously published magnitudes are no longer in the K-band sample. Photometric uncertainties were determined by adding a large number of fake galaxies of known magnitudes to the images, and detecting them in both $\mathcal{R}$ and K bands to mimic our photometry of actual galaxies. This process is described in detail by Shapley et al. (2005b).

4.2.3   Mid-IR Imaging

Two of our fields have been imaged by the Infrared Array Camera (IRAC) on the Spitzer Space Telescope. Images of the Q1700 field were obtained in October 2003 during the "In-Orbit Checkout," and have been previously discussed in detail (Barmby et al. 2004; Shapley et al. 2005b). We also make use of the fully reduced IRAC mosaics of the GOODS-N field, which were made public in the first data release of the GOODS Legacy project (M. Dickinson, PI). These images are described in more detail by Reddy et al. (2005). For both fields we use the photometric procedure described by Shapley et al. (2005b).

4.3   Model SEDs and Stellar Masses

4.3.1   Modeling Procedure

We determine best-fit model SEDs and stellar population parameters for the 93 galaxies for which we have K-band magnitudes. Most of these (87) also have J-band magnitudes, and 35 (in the GOODS-N and Q1700 fields) have been observed with the IRAC camera on the Spitzer Space Telescope. For modeling purposes we correct the K magnitudes for Hα emission; the typical correction is 0.1 mag, but for 4/93 objects (Q2343-BX418, Q2343-BM133, Q1623-BX455, Q1623-BX502), it is $\gtrsim 0.4$ mag. We use a modeling procedure identical to that described in detail by Shapley et al. (2005b), and review the method briefly here.

Photometry is fit with the solar metallicity Bruzual & Charlot (2003) models; as shown by Erb et al. (2005a), solar metallicity is a reasonable approximation for the massive galaxies in the sample, and more typical galaxies have metallicities only slightly less than solar. Although we use the models that employ a Salpeter (1955) IMF, it is well-known that the steep faint-end slope of this IMF below 1 $M_\odot$ overpredicts the M/L and stellar mass by a factor of ~ 2 (e.g., Bell et al. 2003; Renzini 2005). For more accurate comparisons of stellar and dynamical masses we have therefore converted all stellar masses and star formation rates to the Chabrier (2003) IMF by dividing by 1.8 (a change in the faint end of the IMF affects only the inferred total stellar mass and star formation rate, because low mass stars do not provide enough flux in our bandpasses to affect the observed photometry). Note that previous model SEDs of some of the same objects discussed here used the Salpeter IMF (Shapley et al. 2004, 2005b).

The Calzetti et al. (2000) starburst attenuation law is used to account for dust extinction. We employ a variety of simple star formation histories of the form SFR $\propto e^{(-t_{sf}/\tau)}$, with $\tau = 10, 20, 50, 100, 200, 500,$ 1000, 2000, and 5000 Myr, as well as $\tau = \infty$ (i.e., constant star formation, CSF). We also briefly consider more complex two-component models, as described below; but as it is generally difficult to constrain the star formation histories even with simple models, we use only the single-component models for the main analysis. For each galaxy we consider a grid of models with ages ranging from 1 Myr to the age of the universe at the redshift of the galaxy, with extinctions ranging from $E(B-V) = 0.0$ to $E(B-V) = 0.7$, and with each of the values of τ listed above. We compare the U_nGRJK+IRAC magnitudes of each model (shifted to the redshift of the galaxy) with the observed photometry, and determine the normalization that minimizes the value of χ^2 with respect to the observed photometry. Thus the reddening and age are determined by the model, and the star formation rate and total stellar mass by the normalization. In most cases all or most values of τ give acceptable values of χ^2, while the constant star formation models give the best agreement with star formation rates from other indicators (see §4.5). For these reasons we adopt the CSF model unless it is a significantly poorer fit than the τ models. CSF models are used for 74 of the 93 objects modeled, and τ models for the remaining 19; as discussed by Shapley et al. (2005b), the most massive galaxies, with $M_\star \gtrsim 10^{11}$ $M_\odot$, are usually significantly better fit by declining models with $\tau = 1$ or 2 Gyr. We discuss this issue further in §4.5. The adopted parameters and values of τ may be found in Table 4.3.

Parameters derived from this type of modeling are subject to substantial degeneracies and systematic uncertainties, primarily because of the difficulty in constraining the star formation history. These issues have been discussed in detail elsewhere (Papovich et al. 2001; Shapley et al. 2001, 2005b). Here we simply note that the primary degeneracy is between age and reddening, and that the stellar mass is less sensitive to the assumed star formation history and thus more tightly constrained. We examine the effects of the assumed star formation history, and of photometric errors, through a series of Monte Carlo simulations that we use to determine uncertainties on the fitted parameters. The simulations are conducted as described by Shapley et al. (2005b); they perturb the observed colors of the galaxy by an amount consistent with the photometric errors, and determine the best-fit SED as usual to the perturbed colors, including the star formation history τ as a free parameter. We conduct 10,000 trials per object to estimate the uncertainties in each fitted parameter, and iteratively determine the mean and standard deviation of each parameter for each object, using 3σ rejection. The resulting mean fractional uncertainties are $\langle \sigma_x/\langle x \rangle \rangle = 0.7, 0.5,$ 0.6, and 0.4 in $E(B-V)$, age, SFR, and stellar mass, respectively. The distributions of $\sigma_x/\langle x \rangle$ are shown in Figure 4.1. The addition of the mid-IR IRAC data significantly improves our determination of stellar masses; for those objects with IRAC data (all those in the Q1700 and HDF fields) we find $\sigma_{M_\star}/M_\star \sim 0.2$, while those without the IRAC data have $\sigma_{M_\star}/M_\star \sim 0.5$. The IRAC data also reduce the uncertainties in the other parameters, though not by as large an amount. Shapley et al. (2005b) show that while the IRAC data significantly reduce the uncertainties in stellar mass, the lack of these data does not substantially change or bias the inferred mass itself, especially when we are able to correct the K-band magnitude for Hα emission.

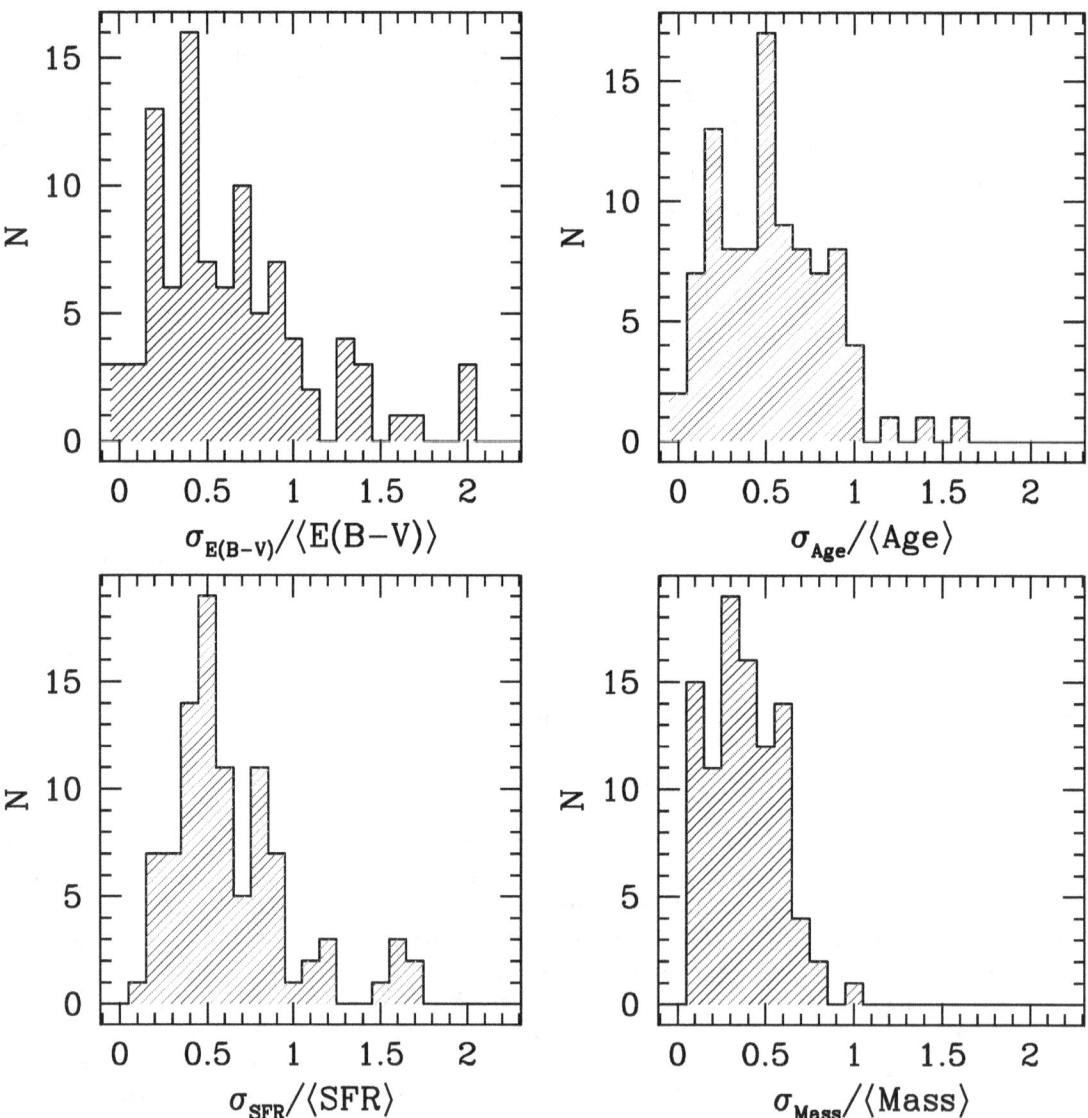

Figure 4.1 The distribution of uncertainties of each parameter fitted by our SED modeling code, from the Monte Carlo simulations described in the text. We show histograms of the ratio of the standard deviation to the mean of each parameter for all the objects modeled, determined iteratively with 3σ rejection. From left to right and top to bottom, we show $E(B-V)$, age, star formation rate, and stellar mass. Stellar masses are the best-constrained parameter, followed by the age of the star formation episode.

Confidence interval plots for $M_\star$ vs. $E(B - V)$ and age vs. $E(B - V)$ for selected objects are shown in Figure 4.2, where the dark contours represent 68% confidence levels and the light 95%. The red $\times$ marks our adopted best fit. These objects are generally representative of the range of properties spanned by the NIRSPEC sample. The first two objects, Q1623-BX502 and Q2343-BX493, have low stellar masses, young ages, and relatively high values of $E(B - V)$. The third galaxy, Q1700-BX536, has properties close to the average of the sample, and also shows how much smaller the confidence intervals can become with the addition of the mid-IR IRAC data. The last object, Q2343-BX610, is a massive, old galaxy that can only be fit with $M_\star > 10^{11}$ M$_\odot$ and age $t_{sf} > 1$ Gyr.

4.3.2 Model SED Results

The best-fit parameters from the SED modeling are given in Table 4.3, and their distributions shown in the histograms in Figure 4.3. The mean stellar mass is 3.6×10^{10} M$_\odot$, and the median is 1.9×10^{10} M$_\odot$. The mean age is 1046 Myr, and the median age is 570 Myr. The sample has a mean $E(B - V)$ of 0.16 and a median of 0.15. The mean SFR is 52 M$_\odot$ yr^{-1}, while the median is 23 M$_\odot$ yr^{-1}; the difference between the two reflects the fact that a few objects are best fit with high SFRs (> 300 M$_\odot$ yr^{-1}). We discuss the results of the models further in the following sections, where we compare them with properties determined from the Hα spectra.

The 93 galaxies whose SEDs we discuss here are drawn from a total sample of 461 objects with model SEDs. This larger sample will be discussed in full elsewhere (Erb et al. 2005, in preparation); for the moment we compare the distribution of stellar masses of the NIRSPEC galaxies with that of the full sample, which is representative of the UV-selected sample except that it excludes the $\sim 20\%$ of objects that are not detected to $K \sim 22.5$ (as we discuss in §4.5, these are likely to be objects with low stellar masses and relatively low star formation rates). For the purposes of comparison with the full sample, we use models for the NIRSPEC galaxies in which we have not corrected the K magnitude for Hα emission (since we do not have Hα fluxes for the full sample); everywhere else in this paper, we use corrected K magnitudes for the modeling as described above. We also use CSF models for all galaxies for this comparison. As shown in Figure 4.4, the subsample for which we have Hα spectra has a slightly higher average stellar mass (3.6×10^{10} M$_\odot$) than that of the UV-selected sample as a whole (2.9×10^{10} M$_\odot$).

4.3.3 Two-Component Models and Maximum Stellar Masses

One limitation of this type of modeling is that it is sensitive to only the current episode of star formation; it is possible for the light from an older, underlying burst to be completely obscured by current star formation (Papovich et al. 2001; Shapley et al. 2005b). Although we can only weakly constrain the star formation histories of individual galaxies even with simple declining models, more complex two-component models are a useful tool to assess how much mass we could plausibly be missing. We use two types of two-component models to examine this question. The first is a maximal mass model, in which we first fit the K-band (and IRAC, when it exists) data with a nearly instantaneous ($\tau = 10$ Myr), maximally old ($z_{form} = 1000$) burst, subtract this model from the observed data, and fit a young model to the (primarily UV) residuals. Such models produce stellar masses ~ 3 times higher than the single-component models on average, and for galaxies with (single-component) $M_\star \lesssim 10^{10}$ M$_\odot$, the two-component total masses are 10–30 times larger. Such models are poorer fits to the data, however, and are not plausible on average because the young components require extreme star formation rates; the average SFR for the young component of these models is ~ 900 M$_\odot$ yr^{-1}, $\gtrsim 30$ times higher than the average star formation rate predicted by X-ray stacking analyses (Reddy & Steidel 2004), and far higher than SFRs from Hα and UV emission (see §4.5). Nevertheless, we cannot rule out the possibility that such models may be correct for some small fraction of the sample.

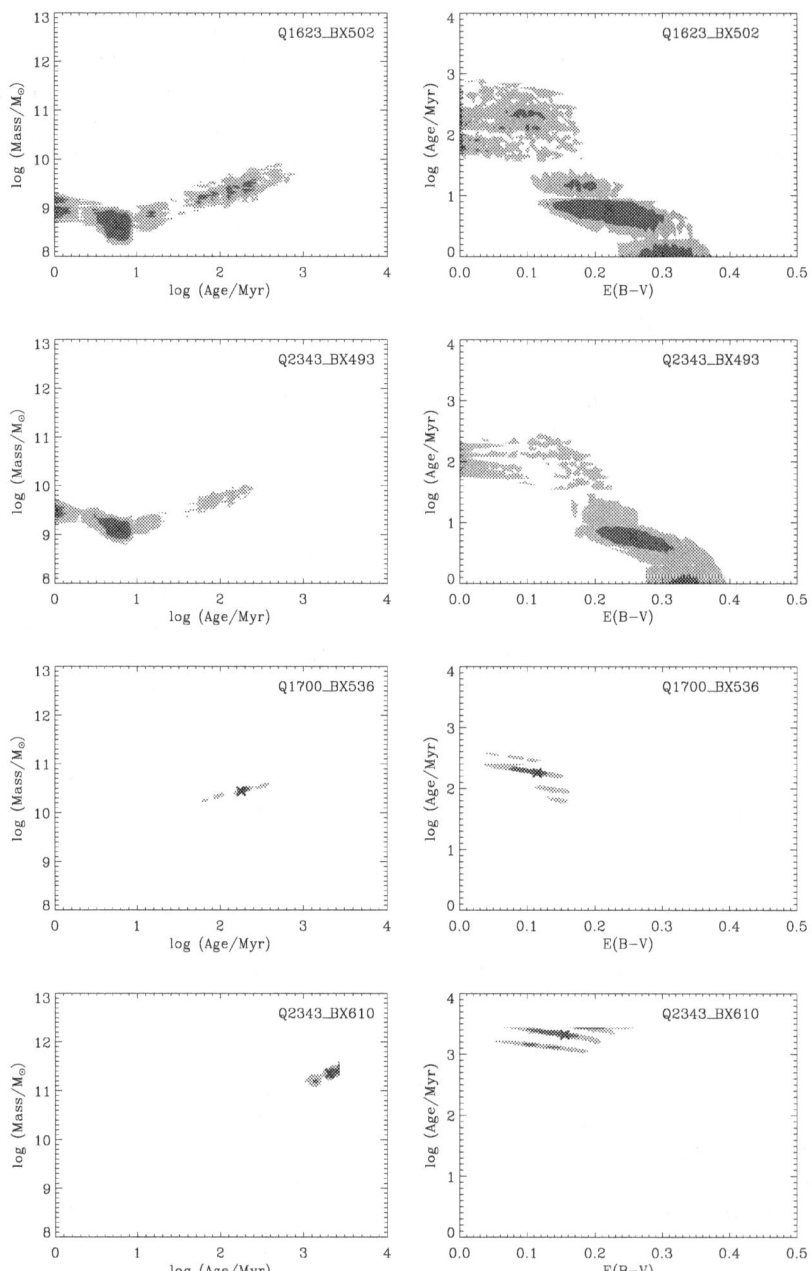

Figure 4.2 Sample confidence intervals for the fitted parameters, for four galaxies in our sample. We show contours of stellar mass vs. age on the left, and age vs. $E(B - V)$ on the right. The dark and light blue regions represent 68% and 95% confidence intervals, respectively, and the red × marks our adopted best fit. The top two objects are characteristic of the young, low stellar mass galaxies in our sample; the third row shows a typical galaxy, with fit very well-constrained by the addition of mid-IR IRAC data; and the bottom panel shows one of the most massive galaxies in the sample, which can only be fit by an old stellar population.

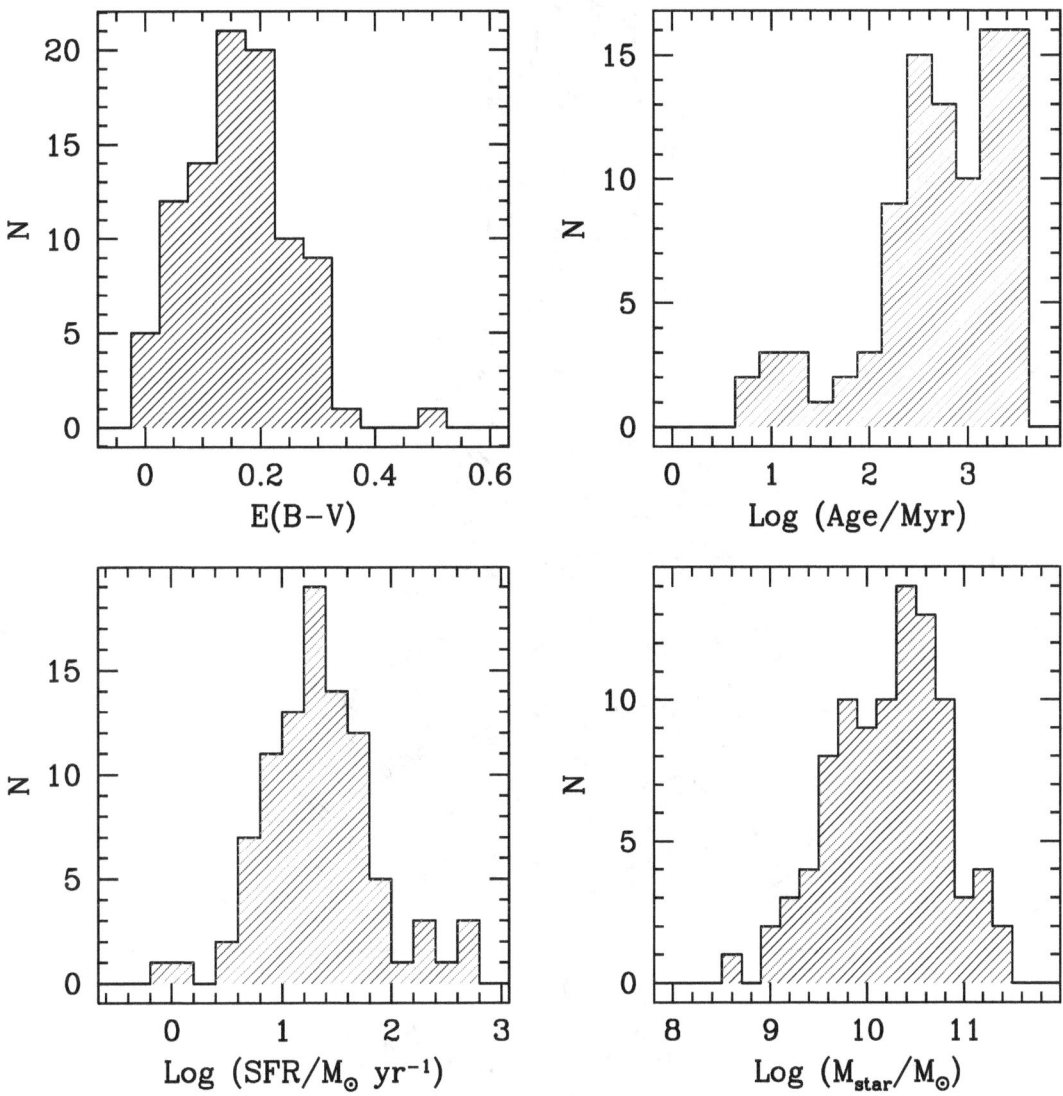

Figure 4.3 Histograms showing the distributions of the results of the SED modeling. From left to right and top to bottom, we show $E(B-V)$, age, star formation rate, and stellar mass. Statistics of the distributions are given in the text.

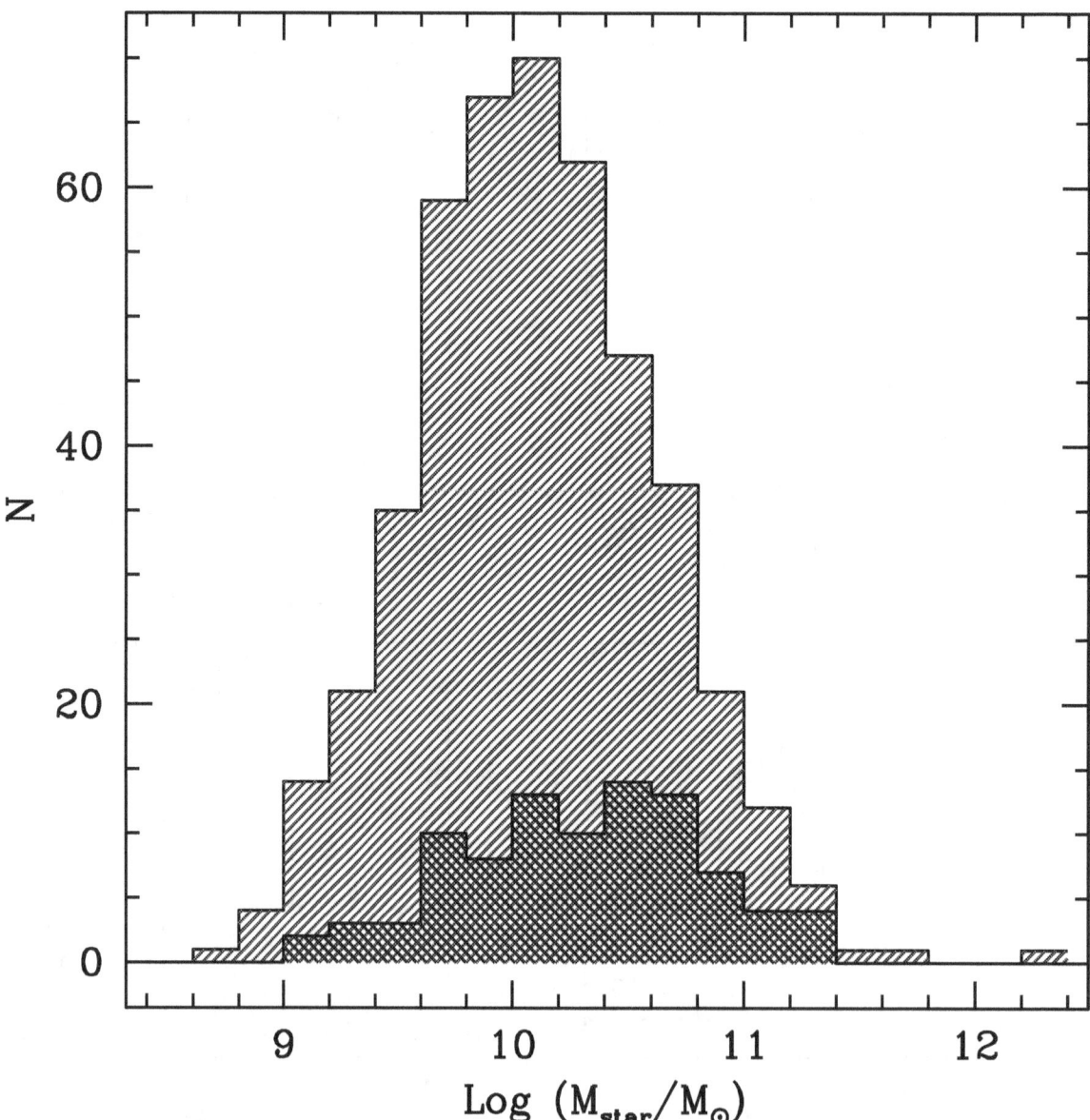

Figure 4.4 A comparison of the stellar masses of the full sample of 461 UV-selected galaxies (red) and the 93 galaxies for which we have model SEDs and Hα spectra (blue). The mean stellar mass of the full sample is 2.9×10^{10} M$_\odot$, while that of the Hα subsample is 3.7×10^{10} M$_\odot$; the NIRSPEC sample is slightly more massive on average because we preferentially targeted some galaxies with red $\mathcal{R} - K$ or $J - K$ colors or bright K magnitudes. For the purposes of this comparison we use constant star formation models for all objects, and the K magnitudes of the NIRSPEC sample have not been corrected for Hα emission.

We also use a more general two-component model, in which we fit a maximally old burst and a current star formation episode (of any age and with any of our values of τ) simultaneously, allowing the total mass to come from any linear combination of the two models. As is expected given the number of free parameters, such models generally fit the data as well as or better than the single-component models. They do not, however, usually significantly increase the total mass. For 40% of the sample, the data is best fit by a current star formation episode alone, with no mass in the old burst; these galaxies span the full range of (single-component) stellar masses, but include almost all of those with the lowest stellar mass. On the other hand, $\sim 20\%$ of the sample is best fit with a model in which the mass in the old component is more than 10 times larger than that produced by the current episode of star formation. These are generally the most massive galaxies in the sample, but because their single-component models are usually already maximally old, their two-component masses are not significantly larger. For the remaining objects, the total mass is split fairly equally between the two components, to within a factor of a few. The average total stellar mass from these models is 1.6 times larger than the average single-component stellar mass, and their other parameters (star formation rate, current star formation age, and reddening) are reasonable. The total two-component mass is more than three times larger than the single component mass in only six cases. In general, then, the data do not favor models with large amounts of mass hidden in old bursts. We discuss both of these models further in the next section, where we compare the stellar and dynamical masses.

4.4 Kinematics

4.4.1 Velocity Dispersions and Dynamical Masses

The most useful kinematic measurement we make is the Hα velocity dispersion σ, because it can be calculated for most of the objects in the sample. Out of the 114 galaxies with Hα detections, we measure a velocity dispersion from the Hα line width for 85 objects. Eleven additional objects have an upper limit on the velocity dispersion, while 12 have FWHM less than the instrumental resolution and six have been rejected due to significant contamination from night sky lines. We calculate the velocity dispersion $\sigma = \mathrm{FWHM}/2.35$, where FWHM is the full width at half maximum after subtraction of the instrumental resolution (15 Å in the K-band) in quadrature. Uncertainties ($\Delta\sigma_{up}$ and $\Delta\sigma_{down}$) are determined by calculating σ_{max} and σ_{min} from the raw line width W and its statistical uncertainty ΔW; $W + \Delta W$ gives σ_{max} and $W - \Delta W$ gives σ_{min}. This results in a larger lower bound on the error when the line width is close to the instrumental resolution, and an upper limit on σ when $(W - \Delta W)^2$ is less than the square of the instrumental resolution. In such cases we give one standard deviation upper limits of $\sigma + \Delta\sigma_{up}$. As we discuss further below, we have also measured the spatial extent of the Hα emission for each object; for 14 objects, the $\mathrm{FWHM}_{\mathrm{H}\alpha}$ is less than the slit width. In such cases the effective resolution is increased by the ratio of the object size to the slit width; we have made this correction in the calculation of σ for these 14 objects. The correction is usually $< 10\%$.

Velocity dispersions are given in column 6 of Table 4.4. From the 85 measurements we find $\langle\sigma\rangle = 120$ km s^{-1}, with a standard deviation of 86 km s^{-1}, while counting the upper limits as detections gives $\langle\sigma\rangle = 112 \pm 85$ km s^{-1}. The observed line widths could be caused by random motions, rotation, or merging, or, most likely, some combination of the above (as we discuss in §4.4.3, we do not believe that galactic winds contribute significantly to the line widths, and the AGN fraction of $\sim 4\%$ means that broadening by AGN is usually not significant). Dynamical masses can be calculated from the line widths via the relation

$$M_{\mathrm{dyn}} = C\sigma^2 r/G \tag{4.1}$$

The factor C depends on the galaxy's mass distribution and velocity field; it can range from < 1 to

~ 10, depending on the mass density profile, the velocity isotropy and relative contributions to σ from random motions or rotation, the assumption of a spherical or disk-like system, and possible differences in distribution between the total mass and that of tracer particles used to measure it. Obviously the definition of the radius r is also crucial. Most of these factors are unknown for the current sample; we adopt a compromise value of $C = 5$ for comparison with previous work (Pettini et al. 2001; Erb et al. 2003, 2004; Shapley et al. 2004; van Dokkum et al. 2004; Swinbank et al. 2004). For the galaxy size r we use the spatial extent of the Hα emission, $r_{\mathrm{H}\alpha} = FWHM_{\mathrm{H}\alpha}/2.35$, after deconvolution of the seeing. The galaxies are spatially resolved in almost all cases; for those five that are not, we use the smallest size measured under the same seeing conditions as an upper limit. We find a mean and standard deviation $\langle r_{\mathrm{H}\alpha} \rangle = 0.3 \pm 0.1''$. We can compare this with rest-frame UV sizes obtained from the high resolution ACS images of the GOODS-N field, where $\langle r \rangle = 0.26''$; this is the mean rms of the region containing half the light, so we expect it to be somewhat smaller than our Hα sizes which consider the full extent of the emission. Because it is difficult to continuously monitor the seeing while observing with NIRSPEC, some uncertainty is introduced by our inexact knowledge of the seeing during each observation. This uncertainty is typically 0.1–0.2$''$, but for those objects that are near the resolution limit it may be much larger. Because of this issue, the sizes are uncertain by a factor of ~ 2. Values of $r_{\mathrm{H}\alpha}$ are given in column 5 of Table 4.4, and dynamical masses in column 8. Using the 85 galaxies with well-determined σ, we find a mean dynamical mass $\langle M_{\mathrm{dyn}} \rangle = 7.4 \times 10^{10}$ M$_\odot$. The two largest dynamical masses in the sample are $M_{\mathrm{dyn}} = 3.5 \times 10^{11}$ M$_\odot$ (Q1700-MD174) and $M_{\mathrm{dyn}} = 2.4 \times 10^{12}$ M$_\odot$ (Q1700-MD94); both of these objects are AGN, and their broad Hα lines are therefore not a reflection of the gravitational potential. Neglecting these two objects, and the other AGN that have $M_{\mathrm{dyn}} \sim 4-6 \times 10^{10}$ M$_\odot$ (HDF-BMZ1156, Q1623-BX151 and Q1623-BX663), results in $\langle M_{\mathrm{dyn}} \rangle = 4.3 \times 10^{10}$ M$_\odot$.

Uncertainties in the dynamical masses are considerable. As discussed above, the constant C can vary by an order of magnitude; this includes a wide range of mass and velocity distributions, however, from thin, edge-on disks with pure rotation to isotropic spheres of various concentrations. Although we lack any detailed information, we consider it unlikely that our galaxies span the full range of the distribution, and that it is more likely that the uncertainty in C introduces scatter of perhaps a factor of a few, and some systematic offset as well. Spatially resolved kinematic measurements at high angular resolution will be required to better address this question. Our line widths may also occasionally suffer from contamination by night sky lines; we have removed the most obvious incidents of this, but repeated observations with varying S/N and comparisons of line widths from Hα and [O III]λ5007 (we see no systematic differences between the widths from these two lines) show that this can occasionally affect the line profile by up to $\sim 30\%$ (where we have again removed the most obvious incidents of sky line contamination). Complex spatial and velocity structure can also affect the observed velocity dispersion; as discussed previously (Erb et al. 2003), the (non-AGN) object with the largest velocity dispersion in the sample, Q1623-BX376, is made up of at least two pieces, and the velocity dispersion may reflect the relative velocities of the individual components. This point has recently been emphasized by Colina et al. (2005), who use integral field spectroscopy to study a sample of local ULIRGs. They find that when the center of the galaxy is properly identified the central velocity dispersion of the ionized gas is generally a good tracer of the dynamical mass, but that the peak Hα emission and the largest velocity dispersion do not always correspond to the nucleus. Even when the Hα emission comes from the center of the galaxy, it may not trace the full potential, especially in low S/N spectra of high redshift objects. In some local starburst galaxies, the regions of high surface brightness Hα emission are centrally located, but the emission line widths from this nuclear region do not fully sample the rotation curve (Lehnert & Heckman 1996). Indeed, the mean dynamical mass we derive here is ~ 25 times lower than the typical halo mass as inferred from the galaxies' clustering properties (Adelberger et al. 2005b). A similar situation is found in the Milky Way bulge, which has $\sigma \sim 100$ km s^{-1} (e.g., Tiede & Terndrup 1999; Minniti 1996) and mass $\sim 10^{10}$ M$_\odot$ (Garnett 2002), while residing in a halo of mass $\sim 10^{12}$ M$_\odot$ (Wilkinson & Evans 1999; Zaritsky 1999).

4.4.1.1 Comparisons with Stellar Masses

We turn next to a comparison of the galaxies' dynamical and stellar masses. In Figure 4.5 we plot dynamical mass vs. stellar mass for the 68 galaxies for which we have determined both. The dashed line shows equal masses, while the dotted lines on either side indicate a factor of 7 difference between the two masses (we use a factor of 7 as a rough uncertainty in the $M_{dyn}/M_\star$ ratio resulting from a factor of 2–3 uncertainty in both masses). The mean dynamical mass is ~ 2 times higher than the mean stellar mass. Significant correlation between stellar and dynamical mass is observed in the local universe (e.g., Brinchmann & Ellis 2000), but such a correlation will exist only if the stellar mass makes up a relatively constant fraction of the dynamical mass over the full range of stellar masses. If low stellar mass galaxies have large gas fractions, as is seen in local galaxies (McGaugh & de Blok 1997; Bell & de Jong 2000), they may not have correspondingly small dynamical masses. With these caveats in mind, we perform Spearman and Kendall τ correlation tests and find a probability $P = 0.003$ that the masses are uncorrelated, for a significance of the correlation of 2.9σ. Randomizing the lists of masses to generate 1000 sets of random pairs results in an equally strong correlation for 0.4% of the sets of random pairs. This is consistent with our qualitative expectations from Figure 4.5, where most of the points lie well between the dotted lines marking a factor of 7 difference between the masses (though we see a significant number of outliers with $M_{dyn} \gg M_\star$, to be discussed further below).

Approximately 30% of the galaxies in the sample have $M_\star > M_{dyn}$, clearly an unphysical situation. The discrepancies are usually a factor of 2–3 and almost always less than a factor of ~ 7, which can be easily explained by the factor of a few uncertainty in both masses. There are objects for which the line width is unresolved or for which we are only able to determine an upper limit on the velocity dispersion, and these may have larger $M_\star/M_{dyn}$; two of those with limits on σ result in a limit on the ratio of $M_\star/M_{dyn} > 7$, but both of these are observations with extremely low S/N, making any measurement of σ very difficult. Given that the Hα line widths may not sample the full potential, and that the radius that we use to determine the stellar mass (a photometric isophote that includes all the light) is larger than the radius we use for the dynamical mass, it would not be a surprise to find some objects with $M_\star > M_{dyn}$. If non-isotropic velocities or rotation make significant contributions to σ, projection effects will also lead to low values of σ for some fraction of the sample.

It is potentially more interesting to consider the small but significant fraction of the objects for which $M_{dyn} \gg M_\star$; we focus on the 15% of the sample (10/68) for which $M_{dyn}/M_\star > 7$ (one of these is the AGN Q1700-MD94, which has a dynamical mass of 2.4×10^{12} M$_\odot$ due to its very broad Hα line; this is clearly a problematic estimate of dynamical mass, and we exclude this object in the following discussion). These objects are marked with open diamonds in Figure 4.5, and in many of the subsequent figures as well. The large differences between stellar and dynamical masses suggest the following possibilities: 1) these are galaxies that have recently begun forming stars, and thus have small stellar masses and large gas fractions; or 2) we significantly underestimate the stellar mass for up to $\sim 15\%$ of our sample, presumably because of an undetected old stellar population. We should also consider the possibility that contamination from winds, shock ionization, or AGN causes us to overestimate the dynamical masses of $\sim 15\%$ of the sample. There is no evidence for this third possibility, however; except in the few cases of known AGN, none of the objects show evidence of ionization from any source other than star formation, either from their UV spectra or from high [N II]/Hα ratios. The objects with $M_{dyn} \gg M_\star$ have low stellar masses, and as Erb et al. (2005a) show in the context of estimating metallicity as a function of stellar mass, [N II] is not detected even in a composite spectrum of 15 such objects.

The best-fit ages and Hα equivalent widths (we discuss the equivalent widths more fully in §4.5.1) of the set of galaxies with $M_{dyn}/M_\star > 7$ favor the first possibility. As shown in Figure 4.6, there is a strong correlation between $M_{dyn}/M_\star$ and age, and, with somewhat more scatter, between $M_{dyn}/M_\star$ and $W_{H\alpha}$ (the correlations have 5 and 4 σ significances, respectively). Objects with $M_{dyn} \gg M_\star$ have young best-fit ages,

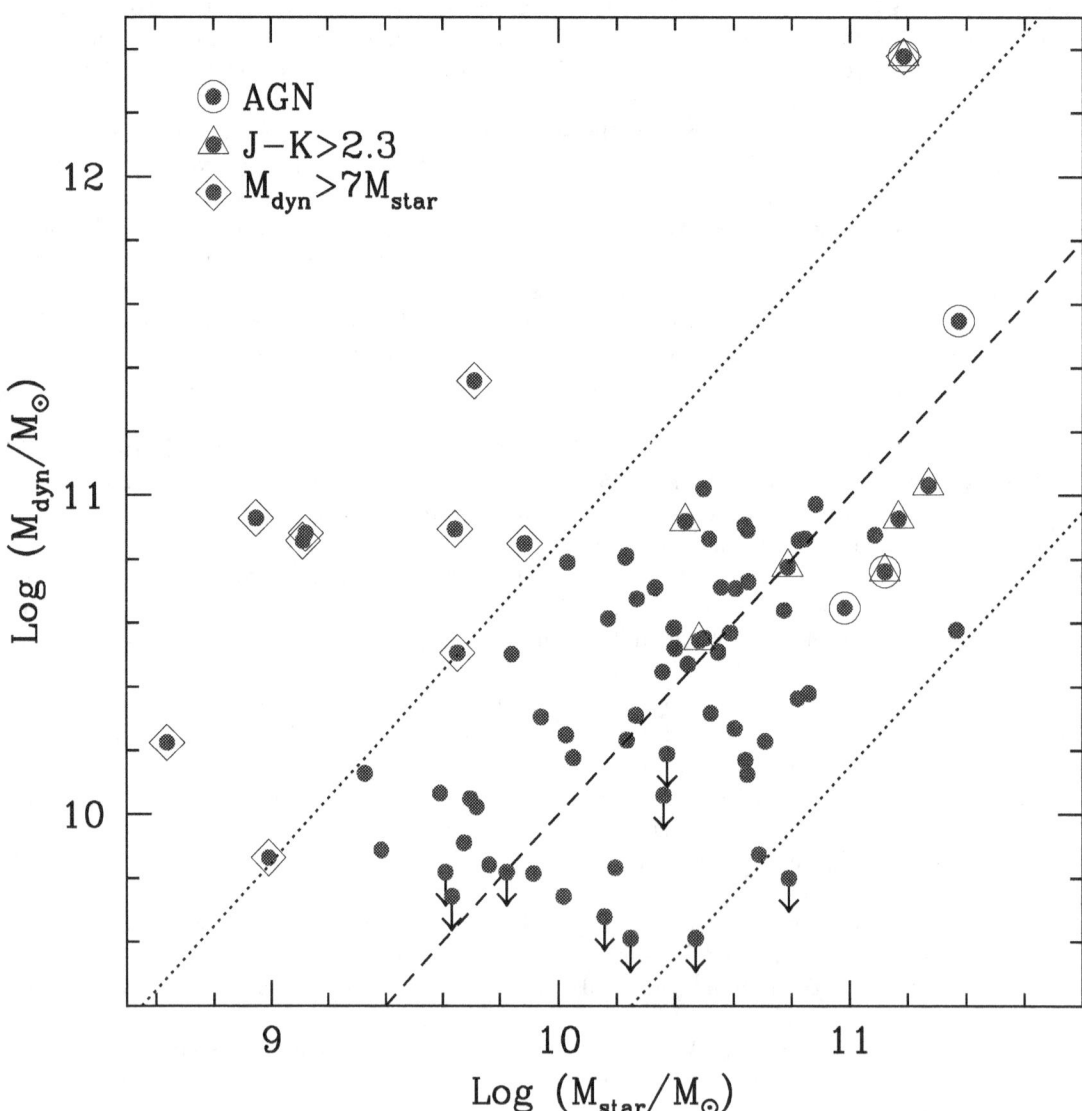

Figure 4.5 Dynamical vs. stellar mass for the 67 galaxies for which we can measure both quantities. We find $\langle M_\star \rangle = 4.1 \times 10^{10}$ M$_\odot$; the mean dynamical mass is $\langle M_{\mathrm{dyn}} \rangle = 8.5 \times 10^{10}$ M$_\odot$. The dashed line shows equal masses, and the dotted lines on either side represent a factor of 7 difference between the masses. Galaxies with $M_{\mathrm{dyn}}/M_\star > 7$ are marked with open diamonds, and we discuss possible reasons for their discrepancy in the text. The masses are correlated with 3 σ significance.

and tend to have high Hα equivalent widths as well. Note that these are not correlations of independent quantities; age and stellar mass are correlated by definition in the constant star formation models we use for the majority of the sample, and the equivalent width depends on the K_s magnitude, which is also closely related to stellar mass. This, combined with the fact that we see less dispersion in dynamical mass than stellar mass, accounts for the correlations in Figure 4.6.

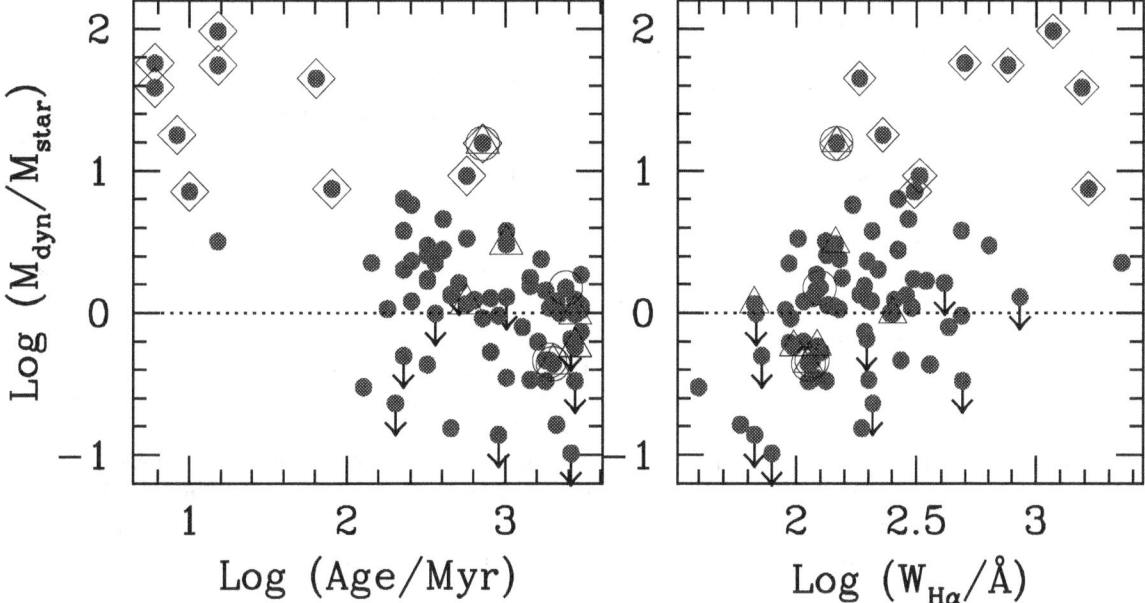

Figure 4.6 Left: The mass ratio $M_{\rm dyn}/M_\star$ vs. the best-fit age from SED modeling, and right, $M_{\rm dyn}/M_\star$ vs. Hα equivalent width $W_{\rm H\alpha}$. Galaxies with high $M_{\rm dyn}/M_\star$ have young ages, and also tend to have high equivalent widths. Symbols are as in Figure 4.5.

We would first like to know whether the young ages of the $M_{\rm dyn} \gg M_\star$ galaxies are significant, given the well-known degeneracies between age and extinction in SED modeling and the considerable uncertainties in the age of a typical object. Confidence intervals for the age–$E(B-V)$ fits for two of the $M_{\rm dyn} \gg M_\star$ galaxies (Q1623-BX502 and Q2343-BX493) are shown in Figure 4.2. These are generally representative of this set of objects. Young ages (and correspondingly high values of $E(B-V)$) are clearly strongly favored, but there is a tail of acceptable solutions extending to higher ages. This set of objects is unique in favoring such young ages. To further assess the significance of the young ages, we divide the sample into four quartiles by $M_{\rm dyn}/M_\star$ (with 16, 17, 17, and 17 galaxies in the quartiles), and perform K-S tests on the age distribution of each quartile. We make use of the Monte Carlo simulations described in §4.3. These simulations perturb the colors of each galaxy according to its photometric errors, and compute the best-fit model for the perturbed colors. We conduct 10,000 trials for each galaxy, and generate lists of the best-fit ages for all trials in each quartile (for 160,000 or 170,000 trial ages in each quartile). We perform a two-sample K-S test on all pairs of the four quartiles, and find that the probability $P \simeq 0$ that the ages in the quartile with the highest $M_{\rm dyn}/M_\star$ are drawn from the same distribution as the remaining quartiles. This test suggests that even though the ages of individual galaxies may not be well-constrained, the younger average age of the subsample with $M_{\rm dyn} \gg M_\star$ is highly significant.

The best-fit ages from the SED fitting represent the age of the current star formation episode, however, not necessarily the galaxy as a whole, which brings us to possibility 2) above: Is there a significant underlying population of old stars in these objects that causes us to underestimate the stellar mass? We address this question through the two-component models described in §4.3.3, which fit the SED with the superposition of a young burst and maximally old population to estimate the maximum stellar mass. The maximal mass models, which fit all the flux redward of the K-band with an old burst, and then fit a young model to the UV residuals, result in stellar masses $\sim$ 5–30 times larger for those galaxies with $M_{\mathrm{dyn}}/M_\star > 7$. If we use the maximal mass models for these objects rather than the single-component models in the comparison with dynamical mass, we find $\langle M_{\mathrm{dyn}}/M_\star \rangle = 2$, as compared to $\langle M_{\mathrm{dyn}}/M_\star \rangle = 35$ using the single-component models. These models are a poorer fit to the data for over 90% of the objects, but provide statistically acceptable fits for $\sim$ 70% of them; four of our 10 galaxies with $M_{\mathrm{dyn}}/M_\star > 7$ are acceptably fit by such a model. Moreover, if no restrictions are applied to the young model, the results are unphysical or implausible. The mean age of the young component is 3.5 Myr, approximately 1/10 the typical dynamical timescale for these objects, while the average SFR is $\sim$ 900 $M_\odot$ yr^{-1}, more than 30 times higher than the average SFR determined for the $z \sim 2$ UV-selected galaxies by X-ray stacking (Reddy & Steidel 2004), and 20–30 times higher than the SFRs we derive from the Hα and UV luminosities (§4.5). Forcing the young component to have a minimum age of 10 Myr also decreases the mean SFR to a more reasonable but still high 100 $M_\odot$ yr^{-1}, but also decreases the quality of the fits such that, for the full sample, less than 40% are acceptable. The more general two-component modeling, in which the total mass comes from a varying linear combination of a maximally old burst and an episode of current star formation, does not favor such extreme models; for the galaxies in question, the use of these more general models increases the average stellar mass by a factor of 3.

More powerful constraints on this question come from the spectra of the galaxies themselves. Using composite spectra of the current sample, Erb et al. (2005a) show that there is a strong trend between metallicity and stellar mass in the $z \sim 2$ sample, such that the lowest stellar mass galaxies also have low metallicities. These objects probably have large gas fractions, accounting for their relatively low metal content (we discuss our indirect derivations of the gas fractions further in §4.5). These results strongly favor the hypothesis that the large dynamical masses of these objects are due to a large gas mass rather than a significant population of old stars. From both considerations of dynamical mass and baryonic mass (gas and stars, §4.5), it seems clear that objects with a best-fit stellar mass of $M_\star \lesssim 10^9$ $M_\odot$ are not in fact significantly less massive than the rest of the sample. This is not an unexpected result given the clustering properties of the $z \sim 2$ galaxies, which imply a minimum halo mass of $\sim 10^{11.8}$ $M_\odot$ (Adelberger et al. 2005b).

4.4.1.2   Velocity Dispersion and Optical Luminosity

Correlations between the optical luminosity L and velocity dispersion σ of local galaxies (the Faber–Jackson relation, $L \propto \sigma^4$; Faber & Jackson 1976), and between their velocity dispersions and black hole masses M_{BH} (Ferrarese & Merritt 2000; Tremaine et al. 2002), are well-known and generally considered to arise from the regulation of star formation and accretion by feedback from star formation and/or AGN activity (e.g., Murray et al. 2005; Di Matteo et al. 2005; Begelman & Nath 2005; Robertson et al. 2005 and references therein). Efforts to find similar relations at high redshift focus on luminosity, stellar mass, and velocity dispersion, as direct measurements of black hole masses are impossible. These studies have generally suffered from small sample sizes and resulted in weak or absent correlations with large amounts of scatter (Pettini et al. 2001; van Dokkum et al. 2004; Swinbank et al. 2004). We revisit the relation between σ and optical luminosity with a sample of 77 $z \sim 2$ UV-selected galaxies, augmented by 21 SCUBA galaxies presented by Swinbank et al. (2004), four of the Distant Red Galaxies (DRGs) discussed by van Dokkum et al. (2004), and 16 Lyman-break galaxies at $z \sim 3$ (Pettini et al. 2001 give velocity dispersions for nine galaxies observed in K and (usually) J by Shapley et al. 2001, and we have subsequently observed seven more). Our total sample thus consists of 118 galaxies with a mean spectroscopic redshift of $\langle z \rangle = 2.38 \pm 0.37$.

The left-hand panel of Figure 4.7 shows the luminosity implied by the rest-frame V-band absolute magnitude, which we determine by multiplying the best-fit SED with the redshifted V transmission curve, plotted against the velocity dispersion σ; none of the luminosities have been corrected for extinction. We have used photometry from Shapley et al. (2001) to remodel the SEDs of the $z \sim 3$ LBGs, for consistency with the $z \sim 2$ sample. The solid line on the left is the maximum luminosity $L_M \simeq (4 f_g c/G)\sigma^4$ proposed by Murray et al. (2005), with the gas fraction $f_g = 1.0$; this is the maximum luminosity a star-forming galaxy can attain before feedback in the form of momentum-driven winds from radiation pressure drives the gas out of the galaxy. The dashed line on the right shows the local i-band[1] Faber-Jackson relation derived from 9000 SDSS galaxies by Bernardi et al. (2003). The two lines bracket most of the data points, but the evidence for correlation is weak; using a Spearman test, we find that the probability that the data are uncorrelated is $P = 0.12$, for a significance of 1.5σ. Repeating the test without the SCUBA galaxies, however, gives a correlation with 3.3σ significance and $P = 0.0007$.

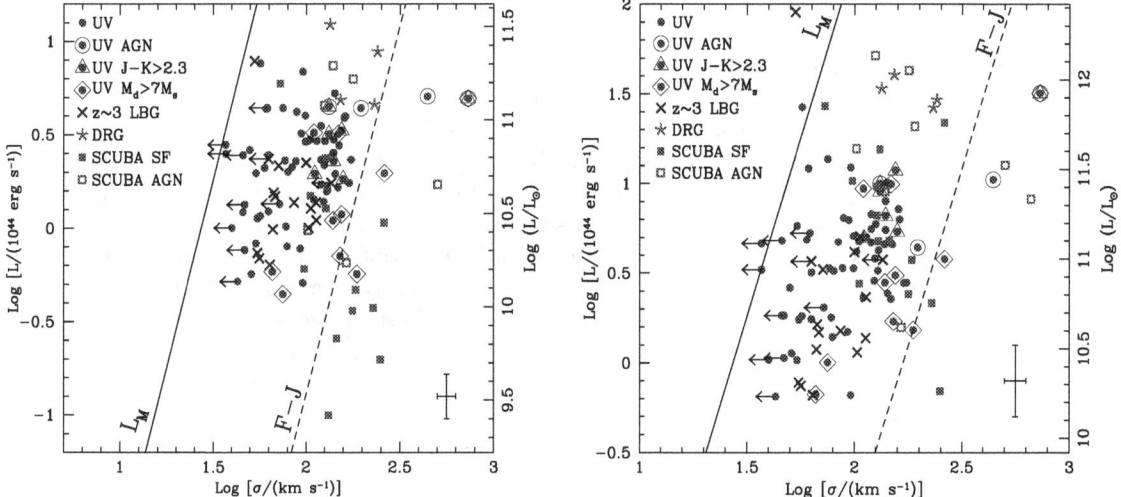

Figure 4.7 Left: Velocity dispersion vs. luminosity calculated from the rest-frame V magnitude, without correcting for extinction. Red circles are the current sample, blue ×s are Lyman break galaxies at $z \sim 3$, green stars are the DRGs presented by van Dokkum et al. (2004), and the purple squares represent the SCUBA galaxies of Swinbank et al. (2004). The dashed line at right shows the local Faber-Jackson relation, and the solid line at left is the maximum wind-driven starburst luminosity of Murray et al. (2005). Right: The same, with luminosities corrected for extinction as described in the text.

The same quantities are shown in the right-hand panel, except that we have corrected the luminosities for extinction using the Calzetti et al. (2000) extinction law and the best-fit values of $E(B - V)$ from the SED modeling. We also use the extinction-corrected magnitudes M_V from Swinbank et al. (2004), and use the values of A_V determined from SED modeling by van Dokkum et al. (2004) to correct their absolute V magnitudes. Again the vast majority of the data points fall between the local Faber-Jackson relation and L_M. There is also an increase in the significance of the correlation; the Spearman and Kendall

---

[1]The slope of the Faber-Jackson relation is the same within the uncertainties in all bands, while the zeropoint decreases slightly at shorter wavelengths.

τ tests give a probability $P = 2 \times 10^{-5}$ that the data are uncorrelated, for a significance of 4.1σ. The slope of the correlation is not well-constrained, and depends strongly on the fitting procedure and the relative uncertainties assumed for L and σ; it is not inconsistent with ~ 4, the observed slope of the local Faber-Jackson relation. While we show the maximum luminosity L_M of Murray et al. (2005) for reference, the rest-frame V-band luminosities plotted here are not the most relevant comparison with L_M. L_M refers to the bolometric luminosity, and even with no dust absorption, a small fraction of the total light from star formation is radiated in the V-band. The fraction depends on the galaxy's age and star formation history, but ranges from ~ 2–10% and is smallest for young objects because massive, luminous stars radiate primarily at shorter wavelengths. The extinction-corrected UV luminosity is therefore a better approximation of the bolometric luminosity for the younger galaxies in our sample; it is up to 10 times higher than the V-band luminosity for those galaxies. If a bolometric correction factor of 10–50 is applied to the luminosities plotted in the right panel of Figure 4.7, the L_M line passes approximately through the center of the distribution.

There is strong evidence for a correlation between L and σ at $z > 2$, but the scatter is considerable and too large to be accounted for by observational errors. σ varies by a factor of ~ 3–4 at a given luminosity, and for most values of σ we see at least an order of magnitude variation in L. We believe that the substantial scatter in the σ–L relation at high redshift is to be expected. The rest-frame optical luminosity is a superposition of light from current star formation and from formed stellar mass, and the star formation rate likely varies on a much shorter timescale than σ. The situation is thus very different from the relatively homogeneous, quiescent early-type galaxies in which the local Faber-Jackson relation is observed, and it would be surprising if we observed a tight correlation. Nevertheless, we have looked for a third parameter that may explain some of the scatter, but without success. For example, the rest-frame optical mass-to-light ratio M/L is a reasonable candidate; as discussed by Shapley et al. (2005b), the varying effects of current star formation and star formation history result in large variation in M/L, typically a factor of ~ 10 and with a maximum variation of a factor of ~ 70. We find nearly identical results when comparing V and $M_\star$ for the present sample. If such variation in M/L were responsible for a significant amount of the scatter in our $\sigma - L$ relation, we would expect that galaxies with the same luminosities but different mass-to-light ratios would fall in roughly different areas of the plot. This is not generally what is found, however. The relationships between M/L, σ, and L are shown in Figure 4.8, where we plot M/L vs. σ and M/L vs. L. M/L depends on L and on σ in much the same way; galaxies with high M/L tend to have both high velocity dispersions and large luminosities, while those with low M/L span the full range in both σ and L. Thus galaxies with the highest M/L tend to occupy a particular region in the $\sigma - L$ plane, but those with lower M/L are found nearly everywhere galaxies are to be found. We have also assessed the $\sigma - L$ scatter as a function of stellar population, but see little connection with age, reddening or star formation rate (although the oldest galaxies also tend to be those with large L and σ discussed above); the uncertainties in these quantities are larger than those in M/L, so such tests are of limited usefulness.

A somewhat less complicated comparison may be between stellar mass and velocity dispersion, quantities that might be expected to be related through the $M_{BH} - \sigma$ relation and the $M_{\text{bulge}} - M_{BH}$ relation (Magorrian et al. 1998; Marconi & Hunt 2003) as well as through the comparison of stellar and dynamical masses discussed above. We plot stellar mass $M_\star$ vs. σ in Figure 4.9; this is similar to Figure 4.5, although here we add the $z \sim 3$ LBGs and the DRGs (stellar masses are not available for the SCUBA galaxies). The correlation has a significance of 3.6σ, about the same as the $\sigma - L$ correlation; the low stellar mass galaxies with large velocity dispersions discussed above are clear outliers (marked with open diamonds). The dotted line is the local relation between bulge stellar mass and velocity dispersion, constructed by combining the M_{bulge}–M_{BH} relation of Marconi & Hunt (2003) and the M_{BH}–σ relation of Tremaine et al. (2002); it provides a surprisingly accurate upper envelope in σ on our plot, as with the exception of AGN and the $M_{\text{dyn}} \gg M_\star$ objects discussed above (marked with open diamonds), virtually all of our points lie on or to the left of this line.

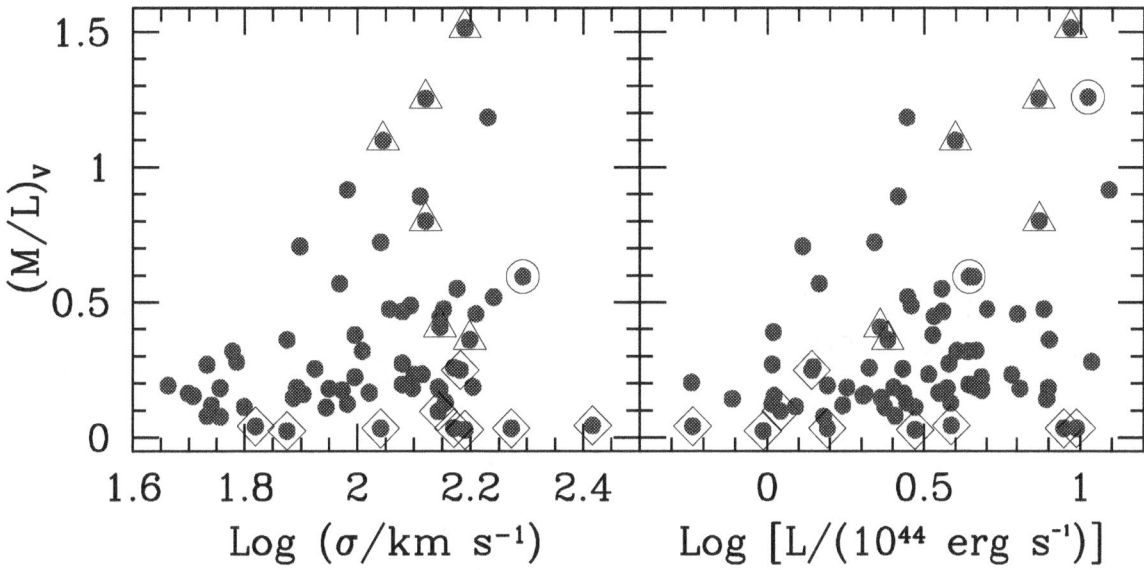

Figure 4.8 Rest-frame V-band mass-to-light ratio M/L vs. velocity dispersion σ (left) and vs. rest-frame optical luminosity (right). Note that objects with low M/L do not depend on σ or L. Symbols are as in Figure 4.5.

At least some theoretical predictions indicate that the correlation between $M_\star$ and σ should be strong at high redshift. The results of one such study are shown by the open symbols, from the numerical simulations of Robertson et al. (2005, see their Figure 3). These simulations test the evolution of the $M_{BH} - \sigma$ relation with redshift using simulations of merging galaxies, incorporating feedback from black hole growth such that $\sim 0.5\%$ of the accreted rest mass energy heats the gas. The $M_{BH} - \sigma$ correlation then results from the regulation of black hole growth by feedback (Di Matteo et al. 2005). The simulation results for $z = 2$ (open squares) and $z = 3$ (open circles) describe the upper envelope in observed σ reasonably well, but the data exhibit considerably more scatter. Given the parameters of the simulations, this is not a surprise; the velocity dispersions in the models (which are the stellar velocity dispersions, not σ from the nebular emission line we use) are measured well after the merger, whereas the data may include a significant fraction of interacting systems, and the simulations are of equal mass mergers with the same orbit for each merger. Both of these conditions probably result in a simulation sample that is considerably more homogeneous than the data set.

A large number of galaxies in our data set have low velocity dispersions for their stellar mass relative to the local correlation (or high stellar mass for their velocity dispersion). In fact this is the opposite of the trend predicted by the simulations, which indicate weak evolution with redshift in the $M_{BH} - \sigma$ and $M_\star - \sigma$ relations, in the sense that velocity dispersions increase at a given stellar mass at higher redshifts (attributed to the steeper potential wells of high redshift galaxies). This difference may reflect the fact that we are probably underestimating the true velocity dispersion in many cases, since we are only sensitive to the highest surface brightness star-forming regions, which may not sample the full potential. It will be interesting to see if deeper spectra with larger telescopes result in a higher average velocity dispersion for this sample, and an accompanying reduction of the scatter. This plot also provides a hint of an evolutionary sequence: If the $M_{dyn}/M_\star > 7$ objects are indeed young galaxies with high gas fractions, they may evolve upward on this plot, significantly increasing their stellar mass with relatively little change in velocity dispersion until they fall on or to the left of the dotted line along with the rest of the sample.

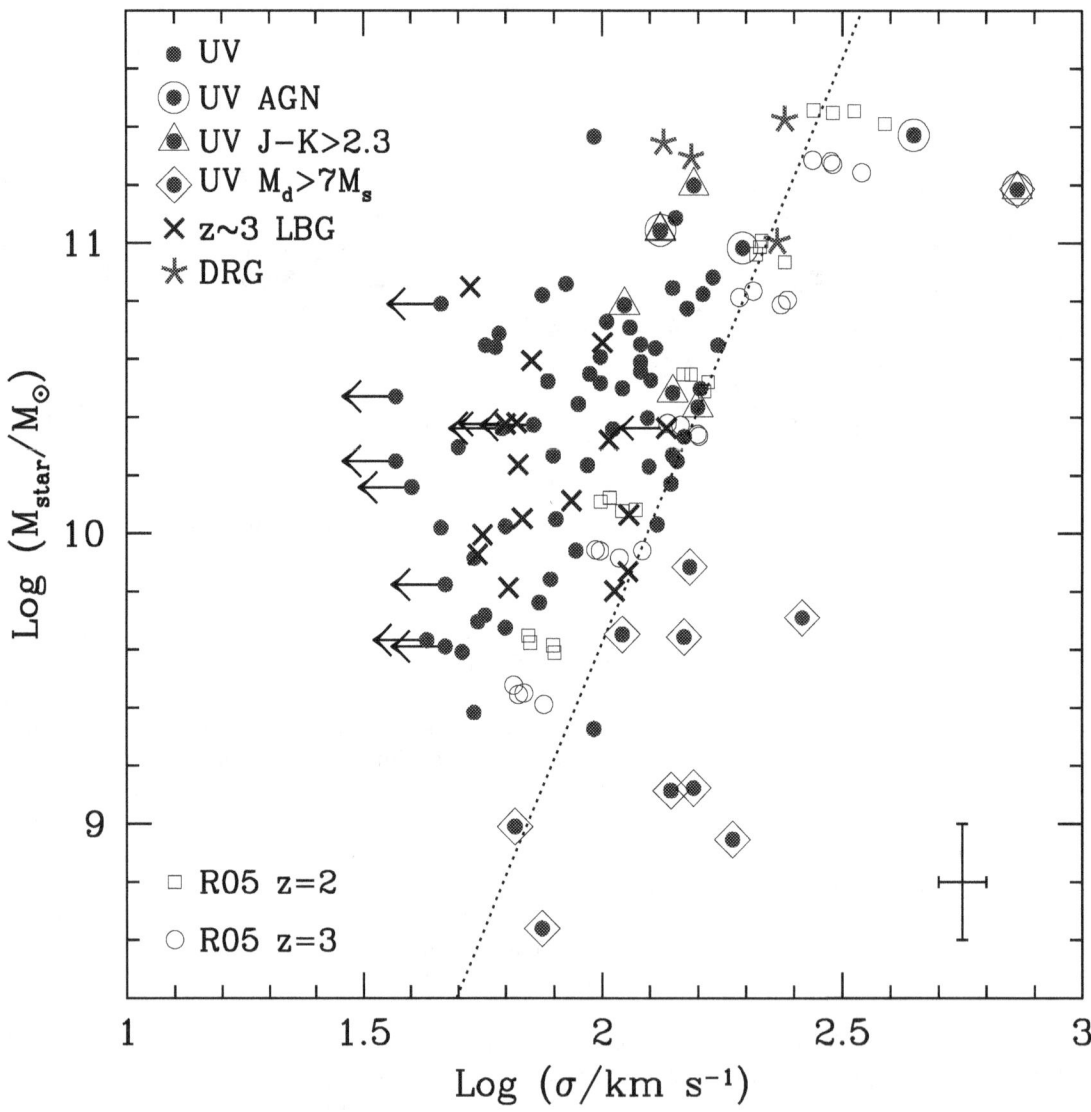

Figure 4.9 A comparison of the velocity dispersion σ and stellar mass $M_\star$ for the samples shown in Figure 4.7 (stellar masses are not available for the SCUBA galaxies). The dotted line shows the local relation between bulge stellar mass and velocity dispersion, constructed by combining the M_{bulge}–M_{BH} relation of Marconi & Hunt (2003) and the M_{BH}–σ relation of Tremaine et al. (2002). Open squares and circles show the results of the simulations of Robertson et al. (2005), which test the evolution of the $M_{BH} - \sigma$ relation with redshift.

Note also that such young objects would probably not be present in the simulations, in which velocity dispersions are measured well after the merger and its accompanying burst of star formation.

4.4.1.3 A Comparison with $z \sim 3$ Lyman-Break Galaxies

It is apparent from Figures 4.7 and 4.9 that the velocity dispersions of the $z \sim 2$ sample and the $z \sim 3$ LBGs have a different distribution. We compare the two directly in Figure 4.10, which shows histograms of the velocity dispersions of the $z \sim 2$ galaxies and the LBGs, after excluding AGN. The mean and the error in the mean of the $z \sim 2$ sample is $\langle \sigma \rangle = 108 \pm 5$ km s^{-1}, while for the $z \sim 3$ LBGs it is $\langle \sigma \rangle = 84 \pm 5$ km s^{-1}. The $z \sim 2$ distribution is significantly broader; objects with low velocity dispersions are common at $z \sim 2$ as well as at $z \sim 3$, but the lower redshift sample contains a large number of objects with $\sigma \gtrsim 130$ km s^{-1}. Only one such galaxy is present in the $z \sim 3$ sample.

Possible selection effects are a concern in attempting to understand this difference. The $z \sim 3$ line widths are measured from [O III]$\lambda 5007$, while Hα is used for the $z \sim 2$ galaxies. To see if systematic differences between line widths measured from the two lines might be responsible for the effect, we have measured line widths from [O III]$\lambda 5007$ for eight of the $z \sim 2$ galaxies that also have Hα line widths. The eight galaxies have Hα velocity dispersions ranging from 60 to 150 km s^{-1}. There is no evidence for a systematic difference in the velocity dispersion from the two lines, with $\langle \sigma_{H\alpha} \rangle = 95$ km s^{-1}, and $\langle \sigma_{[OIII]} \rangle = 97$ km s^{-1}. The individual values of σ usually agree well, and when they do not contamination by sky lines, particularly in the H-band where we measure [O III], appears to be the source of the discrepancy. There is therefore no evidence that the use of different lines is a signifcant issue in the comparison of the two samples.

Another possible factor is the higher redshift of the LBGs. Surface brightness is a strong function of redshift, and it is plausible that we are seeing only brighter, more central regions of the LBGs, and flux from a larger region of the $z \sim 2$ sample. If this is true, we might expect the $z \sim 3$ emission lines to have smaller spatial extents as well as smaller velocity widths. We have measured the spatial extent of 18 of the [O III] emission lines, for comparison with the Hα sizes; as in the $z \sim 2$ sample, most of the lines are spatially resolved, and $\langle r_{[OIII]} \rangle = 0.25'' = 1.9$ kpc, in general agreement with the half-light radii measured from Hubble Space Telescope images by Giavalisco et al. (1996). The mean value of $r_{[OIII]}$ is slightly smaller than the mean Hα spatial extent, $\langle r_{H\alpha} \rangle = 0.30'' = 2.5$ kpc. The difference is not highly significant given the uncertainties involved in correcting for the seeing, and, moreover, it may reflect intrinsic size differences between the samples as well as the effect of surface brightness. Images of the $z \sim 2$ galaxies in the GOODS-N field show that they are often more extended and irregular than the LBGs, and their average size is slightly larger than the average size of LBGs from HST images. This comparison is complicated by selection effects as well. The question of the effect of surface brightness on the line widths remains unresolved; we see no compelling evidence that it affects the $z \sim 3$ line widths, but it cannot be ruled out.

The selection criteria for observation with NIRSPEC were somewhat different for the two samples. The $z \sim 3$ galaxies were chosen primarily because of their UV brightness, with additional objects selected because of their proximity to a QSO sightline. Objects with bright or red near-IR magnitudes or colors were not favored as they occasionally were at $z \sim 2$. If such objects make up most of those with large line widths, this could be a plausible explanation for the differences. An examination of the list of $z \sim 2$ galaxies with $\sigma > 120$ km s^{-1} shows that some of them were indeed selected for such reasons, but an approximately equal number were not. The velocity dispersions of $z \sim 2$ galaxies along QSO sightlines, which should be representative of the sample as a whole, have the same distribution as that of the full $z \sim 2$ sample, indicating that this effect is probably not responsible for the presence of such objects in the $z \sim 2$ sample and their absence among the LBGs. However, whether or not we are able to detect [O III] may depend on the properties of the galaxies. With increasing metallicity, the ratio of [O III]/[O II] decreases, as does the ratio of the [O III] and [O II] lines to Hβ (or Hα) for metallicities above $\sim 1/3 Z_\odot$. Thus the relative fluxes of Hα and [O III] might be expected to depend strongly on metallicity and hence stellar

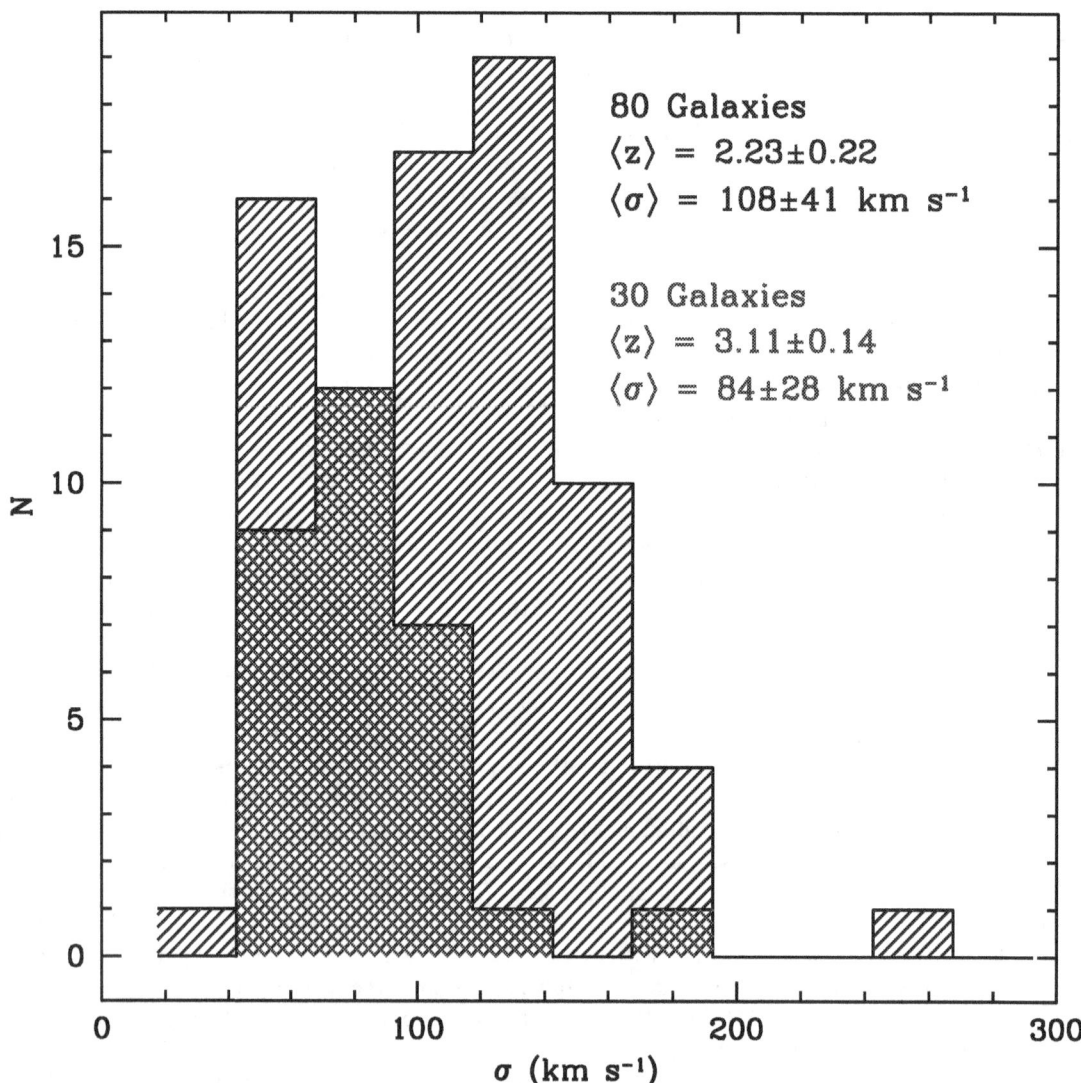

Figure 4.10 The distributions of the velocity dispersions of $z \sim 2$ (larger blue histogram) and $z \sim 3$ (smaller red histogram) galaxies. At $z \sim 2$, σ is derived from the width of the Hα emission line, while [O III] is used for the $z \sim 3$ sample.

mass (Tremonti et al. 2004; Erb et al. 2005a). Although uncertainties in flux calibration make the ratios uncertain, in our sample of $z \sim 2$ galaxies with both Hα and [O III] measurements, $F_{H\alpha}/F_{[OIII]}$ ranges from ~ 0.5 to ~ 5 as stellar mass increases from $\sim 10^9$ to $\sim 10^{11}$ M$_\odot$, and [O III] is not detected for one of the most massive galaxies. It appears that [O III] is weak in massive, relatively metal-rich galaxies, and this could contribute to the absence of objects with large line widths in the $z \sim 3$ NIRSPEC samples.

Not all of the objects with large velocity dispersions in the $z \sim 2$ sample have large stellar masses, however; $\sim 60\%$ of the objects with $M_{\mathrm{dyn}} \gg M_\star$ have $\sigma > 130$ km s^{-1}, and the low stellar mass objects also have low metallicities that should result in strong [O III] lines. If, as we have suggested based on these objects, the velocity dispersion reflects the relatively large mass of the system while the stellar population is still very young, then we might still expect to observe large velocity dispersions in the $z \sim 3$ sample. Such objects are rare, however; young galaxies with $M_{\mathrm{dyn}} \gg M_\star$ and $\sigma \gtrsim 130$ km s^{-1} represent less than 10% of the $z \sim 2$ sample, so finding only one in a sample of 30 objects would not be particularly unexpected (we have no near-IR photometry and no stellar mass estimate for the $z \sim 3$ galaxy with the largest velocity dispersion). It may also be that in many cases the observed velocity dispersion evolves along with the stellar mass, as star formation reaches larger radii with time; perhaps the configuration required to trace a large velocity dispersion with a young stellar population is unusual. Not all of the young, low mass objects in the $z \sim 2$ sample also have large velocity dispersions.

There are reasons to expect physical differences between the $z \sim 2$ and $z \sim 3$ samples. An analysis of their correlation lengths indicates that the $z \sim 2$ galaxies reside in halos ~ 3 times more massive than those hosting the LBGs (Adelberger et al. 2005b). The average stellar mass is also ~ 2 times higher at $z \sim 2$ than at $z \sim 3$, although this comparison is complicated by the different selection techniques of the two samples (Shapley et al. 2005b). On the other hand, the time interval between the mean redshifts of the two samples (~ 870 Myr) is only slightly larger than the median age of the $z \sim 2$ sample, and less than the mean. The most massive galaxies in the $z \sim 2$ sample were most likely already old and massive (with $M_\star \gtrsim 10^{11}$ M$_\odot$) at $z \sim 3$ (Shapley et al. 2005b), and Shapley et al. (2004) show that their progenitors are likely to be found in the LBG sample at $z \sim 3$. In short, while we might expect a lower average velocity dispersion at $z \sim 3$, an almost total absence of objects with large line widths is somewhat surprising.

This question will probably not be resolved until further observations lead to a better understanding of the velocity dispersion at high redshift, and its relation to other properties of the galaxies. In particular, it will be interesting to see if deeper spectra of the $z \sim 3$ galaxies reach a lower surface brightness threshold and result in larger velocity dispersions. For the moment, we conclude that the lack of galaxies with large line widths in the LBG sample is probably due to the smaller fraction of massive galaxies at $z \sim 3$, the rarity of low mass and low metallicity galaxies with large velocity dispersions, and the decrease in [O III] flux with metallicity.

4.4.2 Spatially Resolved Kinematics

In our initial study of 16 Hα spectra of $z \sim 2$ galaxies, nearly 40% of the sample showed spatially resolved and tilted emission lines (Erb et al. 2003). In the current, enlarged sample of 114 objects the fraction is much smaller, 14/114 or 12%; this is probably both because of exceptionally good conditions during our first observing run (we observed six of the 14 objects during this run) and simply because of small number statistics. Velocity shear may be caused by rotation, merging, or some combination of the two, and whether or not it is detected in a nebular emission line depends on the size, surface brightness and velocity structure of the object, its inclination, and the alignment of the slit with respect to the major axis. The seeing during the observations plays a crucial role as well, as Erb et al. (2004) show with repeated observations of the same object. Inclinations and major axes are unknown for nearly all of our sample, as is the primary cause of the velocity shear, and therefore this fraction of 12% does *not* represent the true fraction of rotating or merging objects among the $z \sim 2$ galaxies, though it is a lower limit; moreover, given the limitations imposed by the seeing, the true fraction is not possible to determine.

In order for the tilt of an emission line to be considered significant, we required that the observed spatial extent be at least 1.5 times larger than the seeing disk, and that the peak-to-peak amplitude of the shear be at least 4 times larger than the typical velocity uncertainty. Five of the 14 objects with spatially resolved velocity shear have not been previously discussed (six are described by Erb et al. 2003, two by Erb et al. 2004, and one by Shapley et al. 2004). We plot the observed velocities with respect to systemic as a function of slit position for these five objects in Figure 4.11; we also include Q1623-BX528, the object discussed by Shapley et al. 2004, because such a diagram was not presented in that paper. These diagrams are constructed by stepping along the slit pixel by pixel, summing the flux of each pixel and its neighbor on either side to increase the S/N, and measuring the centroid in velocity at each position. We emphasize that these are not rotation curves of the variety plotted for local galaxies. The points are highly correlated because of the seeing ($\sim 0.5''$ for the observations presented here, thus there are approximately four points per resolution element; we plot one point per pixel to give a clear picture of the *observed* velocity field), and recovery of the true velocity structure requires a model for the structure of the object because of the degeneracies created by the seeing (see Law et al. 2005 for a demonstration). Although the physical scale corresponding to the angular scale is given at the top of each panel for reference, the implied mapping between physical radius and velocity is not a true representation of the structure of the object.

Although the seeing prevents us from determining the true velocity field of the objects, we can at least determine a lower limit on the amplitude of the velocity shear from the uncorrelated endpoints of each curve. We calculate the observed $v_c = (v_{max} - v_{min})/2$ for each object, where v_{max} and v_{min} are with respect to the systemic redshift. This estimate is almost certainly less than the true velocity shear because the inclinations of the galaxies are unknown; in most cases we have made no attempt to align the slit with the major axis; and the seeing effectively reduces the velocities at the endpoints by mixing the light from the edges of the galaxy with emission from higher surface brightness, lower velocity regions toward the center. Even under ideal conditions, the Hα emission may not trace the true circular velocity; Lehnert & Heckman (1996) find that in local starburst galaxies the regions of high Hα surface brightness sample only the inner, solid-body portion of the rotation curve. The observed values of v_c for the 14 objects with velocity shear are given in Table 4.4, and in the left panel of Figure 4.12 they are plotted against the velocity dispersion σ. The dotted line marks equal values; we see that $v_c \sim \sigma$ for most objects, though with considerable scatter.

The relationship between σ and the terminal velocity V_c (we use v_c to refer to our observed velocity shear, and V_c to indicate the true circular velocity of a disk) has been quantified in samples of more local galaxies. Using a sample of faint blue galaxies at $z \sim 0.25$, Rix et al. (1997) find $\sigma \sim 0.6\,V_c$; this is the approximate mean of a broad probability distribution for σ/V_c, accounting for unknown inclinations, asymmetric and centrally concentrated line emission, the shape of the rotation curve, seeing, and other observational effects. More recently, Pizzella et al. (2005) have examined the correlation between the central velocity dispersion of the spheroid and the circular velocity on the flat portion of the rotation curve in a sample of disk and elliptical galaxies. They find $V_c = 1.32\,\sigma + 46$ for elliptical and high surface brightness disk galaxies, for $\sigma \gtrsim 50$ km s^{-1} and with velocities in km s^{-1}.

It is clear that σ and v_c of the galaxies in our sample do not follow these distributions; instead we find a mean $\langle \sigma/v_c \rangle \sim 1.2$ (while it is not possible for V_c to be less than σ, in the limit as the galaxy becomes spatially unresolved the *observed* $v_c \to 0$, and $\sigma/v_c \to \infty$). Because our measurements of σ are less affected by the seeing, they are undoubtably more reliable than our measurements of v_c. If we assume for the moment that our galaxies are rotating disks (which is by no means certain), we can use one of these relations to predict V_c for our sample. This is done in the right panel of Figure 4.12, which shows the observed v_c plotted against the value of V_c predicted from σ using the relation of Pizzella et al. (2005). If the $z \sim 2$ galaxies obey this relation, we underestimate the true circular velocities by an average factor of ~ 2. We have already shown that a change in the seeing from $\sim 0.5''$ to $\sim 0.9''$ can reduce the observed

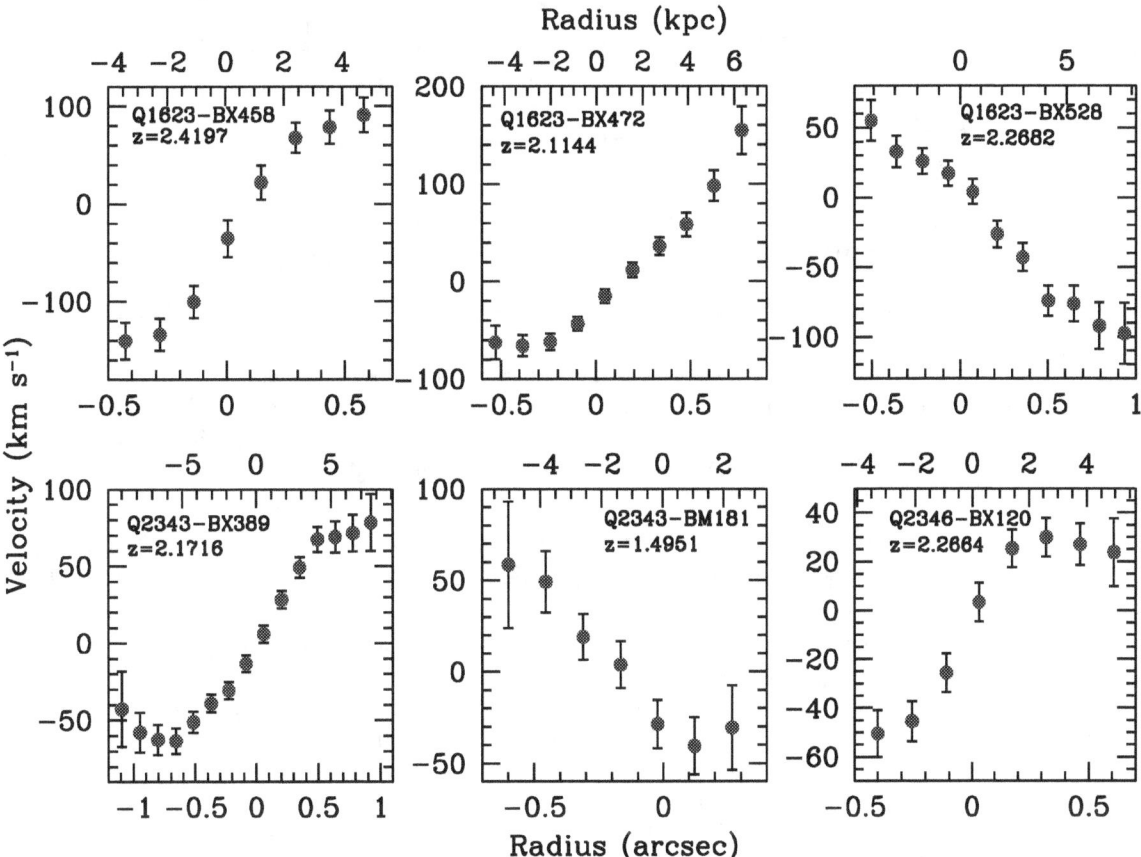

Figure 4.11 Observed velocity as a function of slit position for objects with spatially resolved and tilted Hα emission lines. The seeing for these observations was ∼ 0.5″, so the points shown are highly correlated, with approximately four points per resolution element. We plot one point per pixel to show the observed velocity field clearly, but the blurring of the seeing means that these diagrams do not represent the true velocity structure of the galaxies.

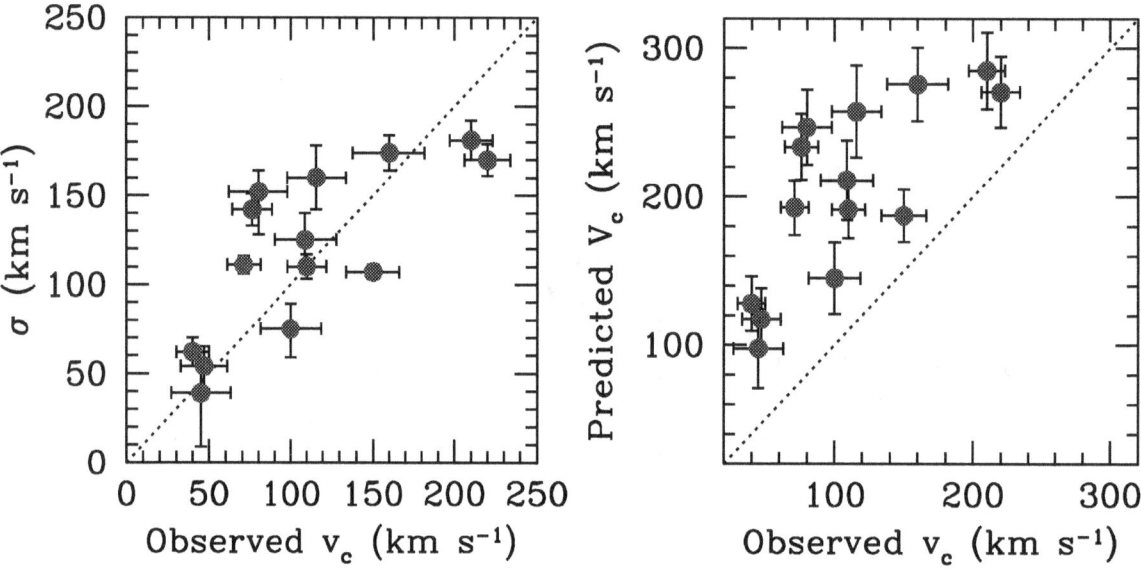

Figure 4.12 Left: A comparison of the velocity dispersion σ and the observed velocity shear $v_c = (v_{\max} - v_{\min})/2$ for the 14 objects with tilted lines. The dotted line shows equal values. Right: The observed velocity shear v_c plotted against the circular velocity V_c predicted from the velocity dispersion using the empirical relation of Pizzella et al. (2005).

v_c by a factor of ~ 2 while changing σ by less than 10% (Erb et al. 2004), so it may not be unreasonable to suppose that a change in resolution from $\sim 0.1''$ or better to $\sim 0.5''$ might have a similar effect. Deep observations of these objects at high angular resolution will be required to obtain a true measure of V_c.

Next we ask whether or not the galaxies that display velocity shear are different from the rest of the galaxies in any significant way. Figure 4.12 shows that galaxies with shear span nearly the full range in velocity dispersion; the mean value of σ for the galaxies with shear is 119 km s^{-1}, while the mean σ of galaxies without shear is 120 km s^{-1} (we do not include objects with limits on σ, for which we would not be able to detect shear). A K-S test shows that the probability that the two samples are drawn from the same distribution is 37%; it appears that we are not more likely to detect shear in galaxies with a large velocity dispersion. Next we consider the stellar population parameters derived from the SED fits for the galaxies with and without shear. We have modeled the SEDs of 10 of the 14 galaxies with shear (the others are in fields not covered by our K-band imaging or, in the case of Q2343-BM181, not detected in K), and we compare these to the remaining 83 galaxies with model fits. There are no significant differences in the star formation rates and values of $E(B-V)$ of galaxies with and without shear, but there is mild evidence that galaxies with shear tend to be older; the median age of the galaxies with shear is 1434 Myr, compared to 509 Myr for the galaxies without shear, and a K-S test finds a probability of 8% that the samples are drawn from the same distribution. The galaxies with shear also have slightly higher stellar masses, with a median of 4.4×10^{10} M$_\odot$ compared to 1.8×10^{10} for those without shear, while a K-S test finds $P = 0.12$. One possible explanation for such trends is that older and more massive galaxies are bigger, and thus more likely to yield a detection of velocity shear. This seems unlikely to be the explanation, however; $r_{H\alpha}$ and age are uncorrelated, while $r_{H\alpha}$ and mass are correlated only weakly. A larger sample of galaxies with shear, and more uniform observing conditions, are required to determine whether or not these differences are significant. Assuming for the moment that they are, they may suggest that the rotation of mature, dynamically relaxed galaxies is a more important contribution to our observed shear than merging, which should not have a preference for older, more massive galaxies. There are also intriguing hints of a connection between velocity shear and the speeds of galactic outflows, as we discuss in the next section.

It appears that nothing in the data is inconsistent with ordered rotation as the primary cause of our observed shear, but the role of merging remains unclear. Given its expected importance at high redshift it is likely that it makes some contribution to our observed velocities. The disentangling of these effects must await high angular resolution spectroscopy with the aid of adaptive optics, particularly integral field spectrographs on 30 m class telescopes; such instruments will be able to map the velocity fields of high redshift galaxies at a resolution impossible with our current data (Law et al. 2005).

4.4.3 Galactic Outflows

It has been known for some time that galactic-scale outflows with velocities of a few hundred km s^{-1} are ubiquitous in star-forming galaxies at $z \sim 2$–3. Evidence is found in the offsets between the redshifts of the nebular emission lines, interstellar absoprtion lines, and Lyα (Pettini et al. 2001), in the relative velocities of stellar, interstellar, and nebular lines in composite UV spectra (Shapley et al. 2003), and in the correlation of C IV systems seen in absorption in QSO spectra with the positions of the galaxies themselves (Adelberger et al. 2003; Adelberger et al. 2005a). The outflowing gas has been observed in detail in the lensed LBG MS1512-cB58; the wind shows a broad range of velocities, with a bulk outflow speed of 255 km s^{-1} and interstellar absorption lines spanning a range of ~ 1000 km s^{-1} (Pettini et al. 2002). Such outflows are also a general feature of starburst galaxies in the local universe, where studies reveal similarly complex velocity stucture and gas in multiple phases (e.g., Heckman et al. 1990; Lehnert & Heckman 1996; Martin 1999; Strickland et al. 2004).

With the current, enlarged sample of $z \sim 2$ galaxies, we re-examine the outflow velocities through a comparison of $z_{\mathrm{H}\alpha}$ and the redshifts of the interstellar absorption lines and Lyα from the rest-frame UV spectra. The redshifts for each galaxy are given in Table 4.4. Absorption redshifts are given only when the spectrum is of sufficient quality to allow a relatively precise measurement (82 of 114 objects), and Lyα redshifts are given for those galaxies with Lyα emission (37 objects). Thirty-six galaxies have all three redshifts. Figure 4.13 shows a histogram of the interstellar and Lyα velocities with respect to the systemic redshift. The distributions have mean $\langle v_{\mathrm{abs}} \rangle = -149$ km s^{-1}, and mean $\langle v_{\mathrm{Ly}\alpha} \rangle = 472$ km s^{-1}, with very little overlap between the two. For objects with both redshifts, $\langle v_{\mathrm{Ly}\alpha} - v_{\mathrm{abs}} \rangle = 645$ km s^{-1}. Typical uncertainties in $v_{\mathrm{Ly}\alpha}$ and v_{abs} are ~ 50 and ~ 100 km s^{-1}, respectively. We interpret the offsets between the redshifts via the standard picture in which we see interstellar absorption from approaching outflowing material, while Lyα is redshifted through resonant scattering off the receding shell on the far side of the galaxy.

It is apparent from Figure 4.13 that a significant fraction of the galaxies have v_{abs} consistent with zero. Given the uncertainties in v_{abs}, this is not particularly significant, but we discuss the issue further because of an intriguing connection with the velocity shear discussed in the previous section. Five of the 15 galaxies with $v_{\mathrm{abs}} \geq 0$ also have Lyα emission, and the mean $\langle v_{\mathrm{Ly}\alpha} \rangle$ of these five objects is 670 km s^{-1}, 200 km s^{-1} higher than the average of the full sample; thus the average full range of the outflow speeds is $\langle v_{\mathrm{Ly}\alpha} - v_{\mathrm{abs}} \rangle = 573$ km s^{-1} for this set of objects, similar to the sample as a whole. This is suggestive of an offset between the redshift of the nebular emission line and the zeropoint of the outflow, rather than of lower outflow speeds. Interestingly, four of the 15 objects with $v_{\mathrm{abs}} \geq 0$ are among those with spatially resolved velocity shear; given our overall detection rate of shear, we would expect to find one or two such objects in a sample of 15. Furthermore, of the 14 objects with tilted emission lines, eight have high quality absorption redshifts, and for these eight, $\langle v_{\mathrm{abs}} \rangle = -47$ km s^{-1}, while 6/8 have $v_{\mathrm{abs}} > -60$ km s^{-1}. The apparent connection between decreased outflow speed and velocity shear can be interpreted in at least two ways. If the velocity shear is indicative of a merger, infalling gas could significantly reduce the average

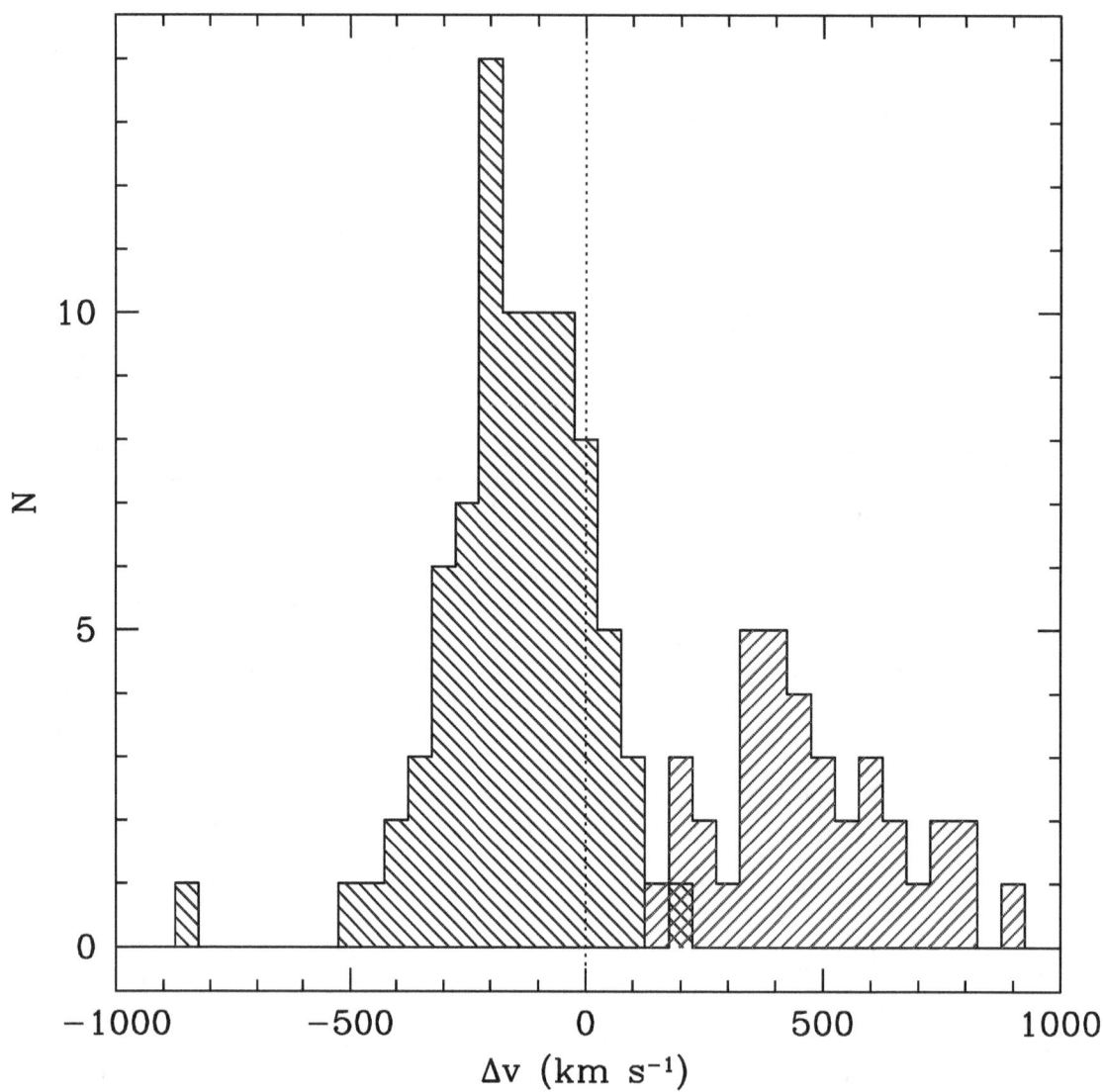

Figure 4.13 The velocities of the UV interstellar absorption lines (blue histogram at left) and Lyα (red histogram at right) with respect to the systemic redshift. We find $\langle v_{\mathrm{abs}} \rangle = -149$ km s^{-1} and $\langle v_{\mathrm{Ly}\alpha} \rangle = 472$ km s^{-1}, with little overlap between the two.

blueshift of the interstellar lines, or the interstellar and nebular redshifts could come from different pieces of the merger. Alternatively, if the tilted emission lines are caused by rotation, the outflows in these objects may be collimated perpendicular to the disk, meaning that projection effects would significantly reduce the observed outflow speed. This would not explain highly redshifted Lyα emission; however, none of the objects with velocity shear also have Lyα in emission. Both effects could be present in our sample, of course, although the significant outflows observed in the vast majority of the sample argue against collimation and projection effects being a major factor in most cases. A better understanding of the causes of the velocity shear will be extremely useful in addressing this question.

Using the dynamical mass from the Hα velocity dispersion, the stellar mass and age determined from the SED fits, and the star formation rate and star formation rate per unit area derived from Hα (see §4.5), we have tested for correlations between galaxy properties and outflow velocity. When each of these quantities is plotted against the outflow velocity (using v_{abs} and $v_{Ly\alpha}$ independently) no significant correlations are found; all of these quantities are consistent with being constant with outflow speed. This lack of correlation is interesting in light of two recent studies relating wind speed to galaxy properties in local starbursts; both Martin (2005) and Rupke et al. (2005) find evidence for correlations of outflow speed with SFR and dynamical mass (represented by the circular velocity). However, a large dynamic range including ∼ 4 orders of magnitude in SFR, and including dwarf starbursts with $v_c \sim 30$ km s^{-1} and SFR < 1 M$_\odot$ yr^{-1}, is needed to detect the trends. The observed correlations flatten for galaxies with SFR $\gtrsim 10$–100 M$_\odot$ yr^{-1} (Rupke et al. 2005), the approximate range of SFRs in our current sample. In other words, no trends are seen when only galaxies with parameters characteristic of our $z \sim 2$ galaxies are considered. An additional issue concerns the outflow velocities themselves. Our values represent the average speed of the outflow, as measured by the centroid of the absorption or emission lines, but it may be the maximum velocity (which presumably reflects the terminal velocity) that is the more fundamental quantity; Rupke et al. (2005) find the strongest correlations with galaxy properties when considering this maximum velocity. Because the interstellar lines are not resolved in our spectra it is not possible to determine the maximum outflow velocity, although the line widths can be used to estimate how much it might be offset from the average. These lines are strongly saturated, indicating that the line widths are due to the velocity dispersion of the absorbing gas rather than the column density of the metals. From a composite UV spectrum of all the galaxies, we find an average equivalent width $W_0 = 1.9$ Å in the C II line at 1334 Å, indicating $\Delta v \gtrsim 430$ km s^{-1}. Thus, on average, the maximum outflow velocities are likely to be at least 200 km s^{-1} higher than the measured v_{abs}.

Finally, we address the question of whether the outflows influence the line widths we use to calculate dynamical masses. In Figure 4.14 we show a comparison of the outflow velocities from Lyα (left) and the interstellar absoprtion lines (right) and the velocity dispersion σ. There is no correlation, indicating that not only are the outflow speeds apparently independent of the dynamical mass of the galaxy, but high outflow speeds do not broaden the lines. As further evidence of the disassociation of the Hα emission and the winds, composite UV spectra constructed by combining individual spectra shifted to the nebular redshift show that Hα is at the same redshift as the stars. Additionally, observations of local galaxies suggest that Hα emission from the outflow would fall far below our detection threshold; Lehnert et al. (1999) study the extended Hα emission from the superwind in the starburst galaxy M82, finding that it has a total luminosity of 2.4×10^{38} ergs s^{-1} and comprises $\sim 0.3\%$ of the total Hα flux. Our typical observed Hα luminosity is 4 orders of magnitude higher than this. Colina et al. (2005) also find, through integral field spectroscopy of Hα emission in local ULIRGs, that the central velocity dispersions are unaffected by outflows.

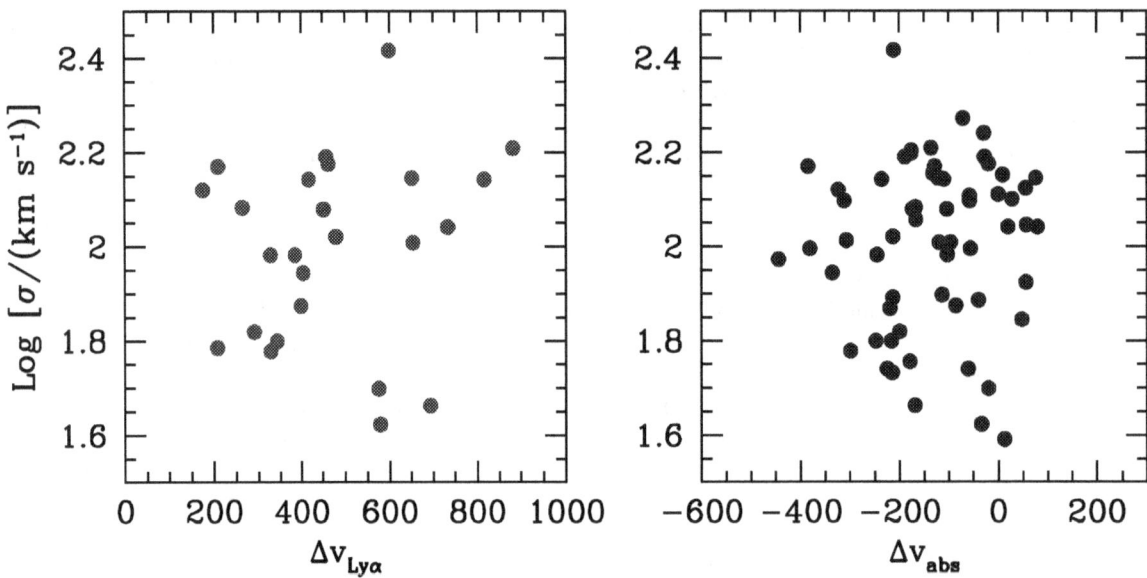

Figure 4.14 Outflow velocities from Lyα (left) and the interstellar absorption lines (right) vs. the velocity dispersion σ. The large outlier Q1623-BX453, with $\Delta v_{\mathrm{abs}} = -870$ km s^{-1} and $\sigma = 61$ km s^{-1}, is not shown. There are no correlations in either case, consistent with our expectation from other arguments that the line widths are not affected by the outflows.

4.5 Star Formation and Extinction

We have three methods of estimating star formation rates for most of the galaxies in the sample: the Hα luminosity, the rest-frame UV continuum, and the normalization of the best-fit model SED (see §4.3). The correspondence of Hα luminosity with SFR in particular is especially useful because it is widely used in the local universe and has recently been studied in detail using large samples of galaxies from the SDSS (Hopkins et al. 2003; Brinchmann et al. 2004). We use the Kennicutt (1998a) transformation between Hα luminosity and SFR, which assumes case B recombination, a Salpeter IMF ranging from 0.1 to 100 M$_\odot$ which we convert to a Chabrier IMF by dividing the SFRs by 1.8, and that all the ionizing photons are reprocessed into nebular line emission. Brinchmann et al. (2004) show that this approximation works well for an average star-forming galaxy, but that massive, metal-rich galaxies produce less Hα luminosity for the same SFR than low mass, metal-poor galaxies. This is probably a metallicity effect, as increased line blanketing in metal-rich stars decreases the number of ionizing photons. Our galaxies follow a trend similar to local galaxies in mass and metallicity, though probably offset to lower metallicities at a given stellar mass (Erb et al. 2005a). The largest dispersion in the conversion factor from Hα luminosity to star formation rate is found for the most massive and metal-rich local galaxies (see Figure 7, Brinchmann et al. 2004); if our sample does not contain galaxies with the highest metallicities observed in the local universe, then the dispersion in the conversion factor is probably less than our uncertainties from other sources, though we may be biased toward overestimating the SFR by ~ 0.1 dex.

In order to calculate SFRs from the UV continuum we use the observed G-band magnitude, which corresponds to a mean rest-frame wavelength of 1480 Å for the galaxies in our sample (except for the five galaxies at $z \sim 1.5$, for which the U_n magnitude corresponds to ~ 1500 Å). We use the Kennicutt (1998a) conversion between 1500 Å luminosity and SFR, which assumes a timescale of $\sim 10^8$ years for the galaxy to reach its full UV continuum luminosity. Because Hα is sensitive to only the most massive stars, it is a more instantaneous measure of SFR than the UV continuum; however, the continuum luminosity rises by a factor of only 1.6 between 10 and 100 Myr, so even for the youngest objects the UV continuum will not severely underestimate the SFR. We again convert the Salpeter IMF to a Chabrier IMF.

We compare the various SFRs in Figure 4.15. The upper left panel shows SFR$_{UV}$ vs. SFR$_{H\alpha}$, without correcting for extinction. There is considerable scatter, but the probability that the data are uncorrelated is $P = 0.0006$, for a significance of the correlation of 3.4σ. We find a mean and standard deviation $\langle SFR_{H\alpha} \rangle = 11 \pm 7$ M$_\odot$ yr^{-1}, and $\langle SFR_{UV} \rangle = 8 \pm 5$ M$_\odot$ yr^{-1}. In the upper right panel a factor of 2 aperture correction has been applied to the Hα fluxes, estimated from narrow-band imaging and the K-band continuum when we detect it in our spectra (see §4.2.1). We have also corrected both fluxes for extinction, using the Calzetti et al. (2000) extinction law and the best-fit values of $E(B - V)$ from the SED fits. For those galaxies that do not have SED fits because we lack the K magnitude, $E(B - V)$ is calculated from the UV continuum slope as measured by the $G - \mathcal{R}$ color, assuming a 570 Myr-old SED with constant star formation; this is the median best-fit age of the current sample. The calculated value of $E(B - V)$ changes by less than 10% for assumed ages from 300–1000 Myr, though for young objects $E(B - V)$ will probably be underestimated using this method. The value of $E(B - V)$ used for each galaxy is shown in Table 4.5; the mean value is $\langle E(B - V) \rangle = 0.16$. We have used the same value of $E(B - V)$ for the stellar UV continuum and for the nebular emission lines, rather than $E(B - V)_{stellar} = 0.4 E(B - V)_{neb}$ as proposed by Calzetti et al. (2000), because the latter assumption significantly overpredicts the Hα SFRs with respect to the UV SFRs. The relative extinction suffered by the stellar continuum and the nebular emission lines is an additional source of uncertainty in our SFRs. After the above corrections, we find $\langle SFR_{H\alpha} \rangle = 31 \pm 18$ M$_\odot$ yr^{-1}, and $\langle SFR_{UV} \rangle = 29 \pm 19$ M$_\odot$ yr^{-1}, using 3σ rejection to compute the statistics in order to prevent the few objects with very high SFRs (particular from the UV luminosity) from biasing the distribution.

The correlation between the corrected Hα and UV SFRs is highly significant (6.8 σ), with an rms scatter of 0.3 dex. Some of this correlation may be due to the extinction correction applied to both SFRs; to test the significance of this effect, we have randomized the lists of uncorrected Hα and UV fluxes to create many sets of mismatched pairs, and applied the same (also randomized) value of $E(B - V)$ to both fluxes in each pair. In 10,000 trials we never observe a correlation as strong as that observed in the real data; the average trial has a correlation significance of 2.8σ induced by the extinction correction. The much higher correlation significance in the real data confirms the underlying correlation of the uncorrected SFRs.

We also compare the corrected Hα SFRs with those determined by the normalization of the best-fitting SED, in the lower panels of Figure 4.15. The lower left panel shows the SFR of our adopted best-fit model vs. SFR$_{H\alpha}$. Again the correlation is strong (5.3σ) and the rms scatter is 0.3 dex. The mean SFR from the SED fits is $\langle SFR_{fit} \rangle = 24 \pm 17$ M$_\odot$ yr^{-1}, again computed with 3σ rejection because of the few objects with very high SFRs. 70% of the objects have SFR$_{H\alpha} >$ SFR$_{fit}$. The points with open circles are those for which we have used declining τ models because they provided a significantly better fit than the constant star formation models; it is clear that the use of τ models depresses the SFR. This can be seen further in the lower right panel of Figure 4.15, in which we plot the SFR of the best-fitting τ model vs. SFR$_{H\alpha}$. The points are coded according to the value of τ: Filled red circles are those galaxies best fit with τ=10, 20, or 50 Myr models, open green circles have τ=100, 200, or 500 Myr, cyan crosses have τ=1, 2, or 5 Gyr, and blue diamonds are constant star formation models. As expected, the steeply declining τ models yield the lowest SFRs, since they allow the SFR to drop significantly during the lifetime of massive stars. The objects with the highest SFRs are also formally best fit by steeply declining models; these are generally young, highly reddened objects that are acceptably fit by all values of τ and have high SFRs for all star formation histories. It is important to bear in mind when considering the τ models that they are undoubtedly an oversimplification of the likely star formation histories. A model with declining star formation may be

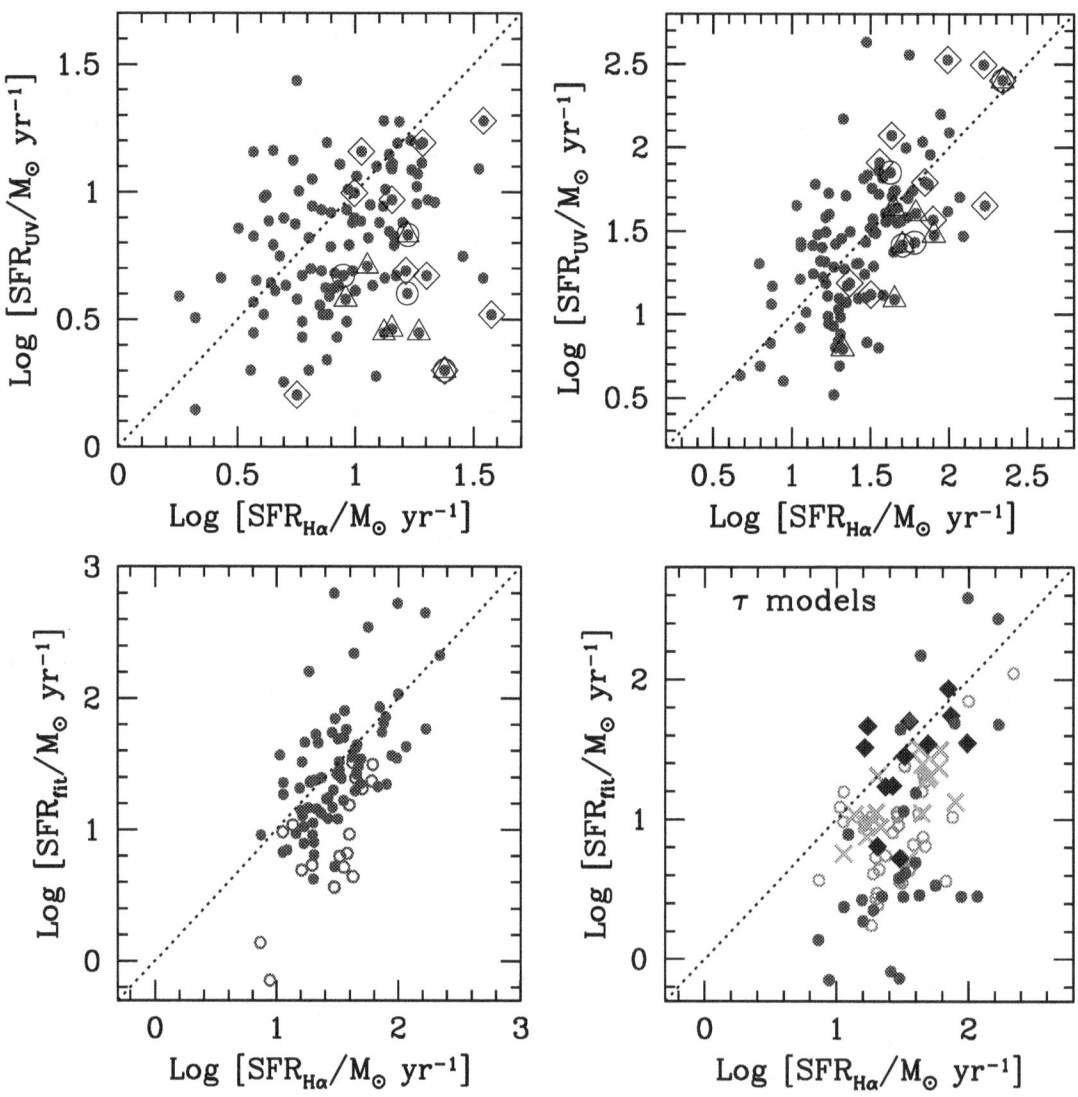

Figure 4.15 A comparison of star formation rates from Hα, the UV continuum, and the SED fits. Upper left: SFR$_{H\alpha}$ vs. SFR$_{UV}$, without correcting for extinction. Symbols are as in Figure 4.5. Upper right: SFR$_{H\alpha}$ vs. SFR$_{UV}$, with both SFRs corrected for extinction and applying a factor of 2 aperture correction to SFR$_{H\alpha}$. Lower left: Corrected SFR$_{H\alpha}$ vs. the SFR obtained from the normalization of the best-fitting SED. Solid symbols are constant star formation models, and the open symbols represent objects for which we have adopted a model with an exponentially decreasing star formation rate. Lower right: Corrected SFR$_{H\alpha}$ vs. the SFR of the best-fitting tau model for each object. Filled red circles are galaxies with τ=10, 20, or 50 Myr, open green circles have τ=100, 200, or 500 Myr, cyan crosses have τ=1, 2, or 5 Gyr, and blue diamonds are constant star formation models. The use of steeply declining τ models decreases the SFR with respect to that found from Hα.

required to obtain an acceptable fit when a galaxy shows significant light from a previous generation of stars as well as a current star formation episode, even if the current episode is best described by constant star formation. In such cases the current SFR is likely to be underestimated. Two-component models that decouple the current star formation episode from the older population are more successful in determining current SFRs; general two-component models (§4.3.3), which add a linear combination of a current episode of contant star formation and an old burst, are significantly better at matching the Hα-determined SFRs of galaxies that require τ models, while still providing an acceptable fit to the SED.

We conclude that a typical galaxy in our sample has a star formation rate of ~ 30 M$_\odot$ yr^{-1}, though the SFRs of individual objects vary by nearly 2 orders of magnitude. The dispersion in the correlations suggests an uncertainty of a factor of ~ 2 for individual galaxies. This result is in very good agreement with the mean SFR of ~ 28 M$_\odot$ yr^{-1} determined for the $z \sim 2$ UV-selected sample from X-ray stacking techniques (Reddy & Steidel 2004; Reddy et al. 2005; we have converted their value to a Chabrier IMF for comparison with our sample).

A further result of the Reddy et al. (2005) study is that the SFR increases with increasing K-band luminosity. We compare the current sample to the results of Reddy et al. (2005) by dividing our sample into bins in K magnitude and finding the average corrected SFR$_{H\alpha}$ in each bin. The results are shown in Figure 4.16, where the red circles are the average Hα SFRs and the blue squares are the SFRs from the X-ray stacking of Reddy et al. (2005). This is a comparison of similar objects, but not the same objects; the X-ray data is only in the GOODS-N field, so the overlap between the two samples is small. The agreement is quite good, although the rise in SFR for K-faint, low stellar mass objects is not seen in the X-ray sample. This is probably a selection effect of our Hα observations; as shown in §4.2, we are less likely to detect Hα emission for objects that are faint in K, unless they have high SFRs. Factoring in non-detections of K-faint galaxies would lower the two right-most points considerably. If the low stellar mass objects in the Hα sample are indeed young starbursts, the relative timescales of X-rays and Hα as star formation rate indicators may also be a factor. The Hα luminosity is nearly instantaneous, while the X-ray luminosity increases for the first $\sim 10^8$ years as O/B stars die and become high-mass X-ray binaries. The X-rays may thus underestimate the SFR for very young objects. In any case, the agreement between the Hα, X-ray, and UV SFRs is quite encouraging. For the remaining analysis, we adopt the corrected Hα SFRs.

4.5.1 Comparisons with Stellar Mass and Star Formation Timescales

The left panel of Figure 4.17 shows the extinction- and aperture-corrected SFR$_{H\alpha}$ plotted against stellar mass. The points marked with open diamonds are objects with $M_{\rm dyn}/M_\star > 7$ (see §4.4.1). There is a general trend in the sense that objects with higher stellar masses have larger SFRs, and the full sample is weakly correlated with a significance of 2.3σ. The objects with high $M_{\rm dyn}/M_\star$ clearly stand out as having anomalously high SFRs for their stellar masses (the object at the upper right is the AGN Q1700-MD94); with these removed from the sample, the significance of the SFR and $M_\star$ correlation is 4σ, with rms scatter of ~ 0.2 dex. To check that this is not caused by the extinction correction, in the right panel of Figure 4.17 we plot the uncorrected SFR$_{H\alpha}$ vs. stellar mass, two entirely independently derived quantities; the plot is very similar, with the $M_{\rm dyn}/M_\star > 7$ objects clear outliers and a general trend of increasing SFR with increasing stellar mass. As we have discussed previously, the $M_{\rm dyn} \gg M_\star$ objects have the youngest ages in the sample, and their Hα SFRs therefore suggest that they are young starbursts. Such a trend between SFR and stellar mass is expected given the correlation between SFR and K magnitude already shown, and the strong correlation between stellar mass and K-band luminosity. Similar trends of increasing SFR with stellar mass are observed in the local universe, with the notable difference that a turnover is observed at $M_\star \gtrsim 3 \times 10^{10}$ M$_\odot$ as massive galaxies tend to be early-type objects that form few stars (Brinchmann et al. 2004).

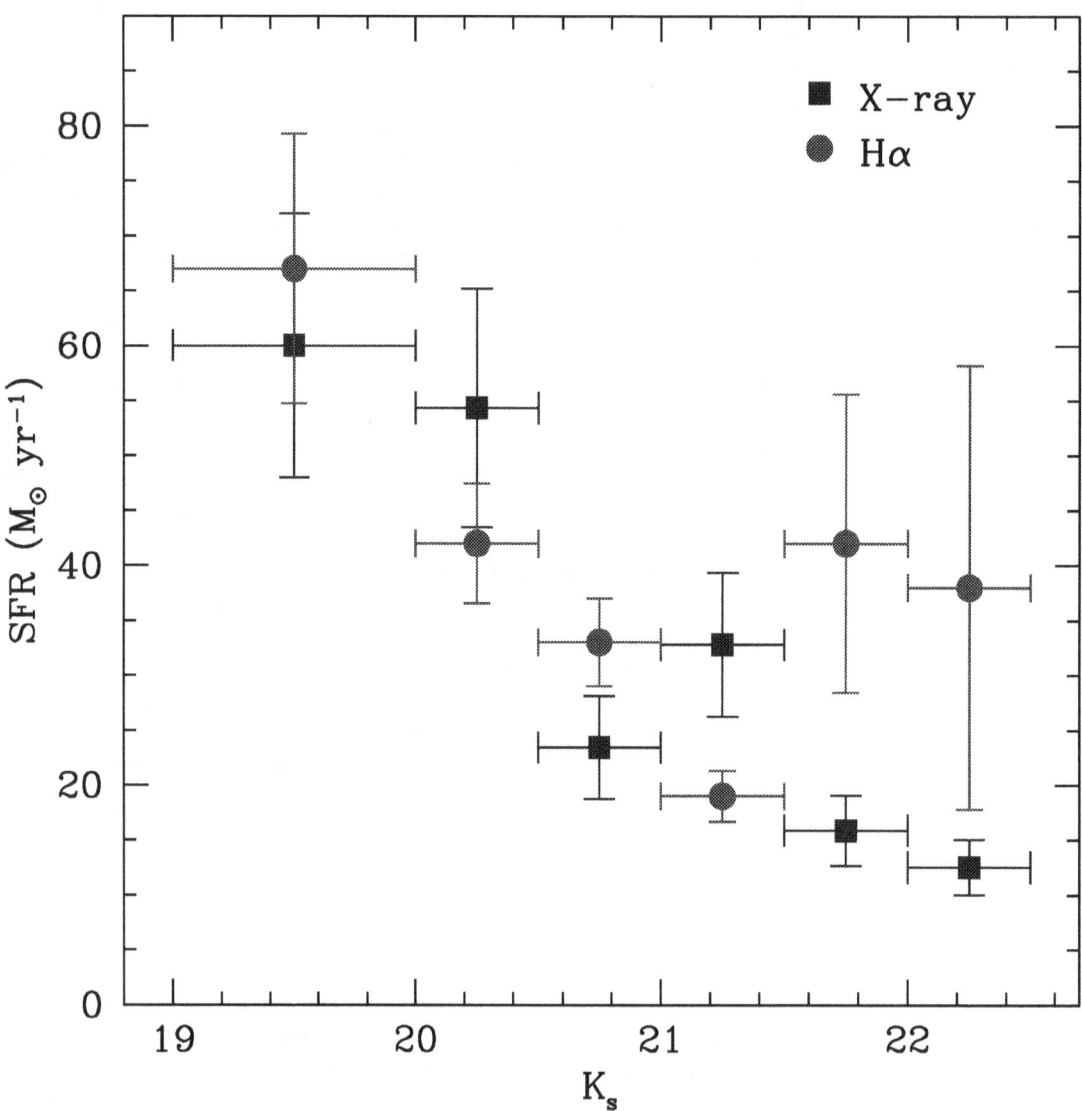

Figure 4.16 Star formation rates from Hα and X-ray stacking, as a function of K magnitude. Red circles from left to right represent the average corrected $\mathrm{SFR_{H\alpha}}$ of galaxies with $19 < K_s \leq 20$ and in 0.5 magnitude bins between $K_s = 20$ and $K_s = 22.5$. The average SFRs determined by stacking deep X-ray images of $z \sim 2$ galaxies in the GOODS-N field in the same ranges of K magnitude are shown by the blue squares (Reddy et al. 2005). The upturn in $\mathrm{SFR_{H\alpha}}$ at faint K magnitudes is a selection effect, because we are less likely to detect Hα in galaxies faint in K.

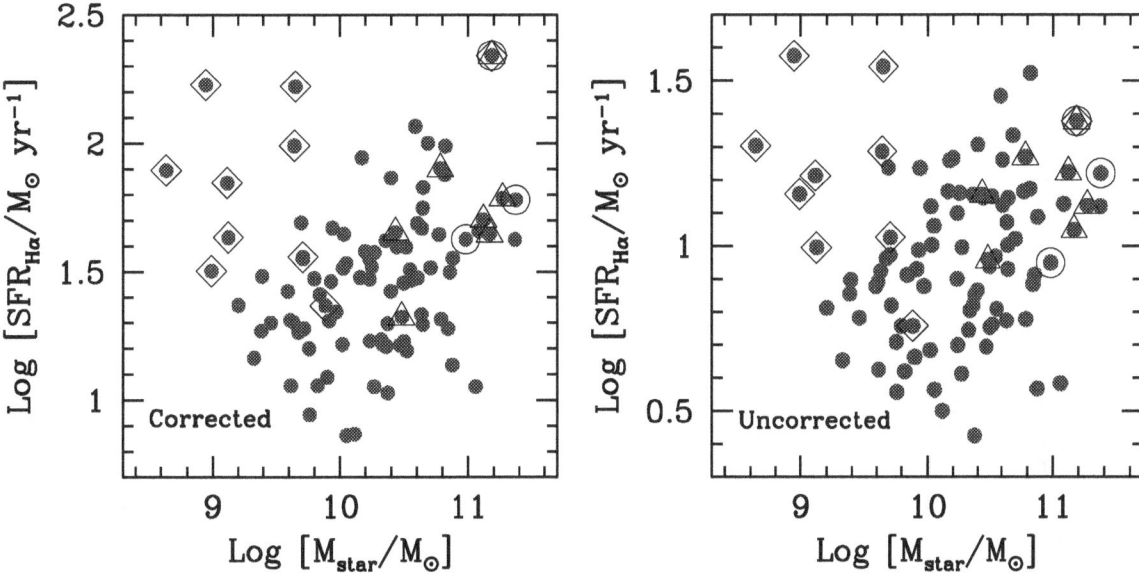

Figure 4.17 Star formation rate from Hα vs. stellar mass, with the SFR corrected for extinction and slit losses at left, and uncorrected at right. In both cases SFR increases with increasing stellar mass, except for the galaxies with $M_{\rm dyn}/M_\star$ (marked with open diamonds). The absence of low mass galaxies with low SFRs is probably a selection effect, as such objects are less likely to be detected both in our K-band images and in Hα. Symbols are as in Figure 4.5.

The SFR and the stellar mass imply a star formation timescale, $T_{\rm SFR} = M_\star/{\rm SFR}$; this is the time required for the galaxy to form all its stellar mass at the current SFR. By comparison with the age of the universe at the redshift of the galaxy and with the inferred age from the SED fits, we may obtain some constraints on the star formation histories. We make these comparisons in Figure 4.18, where in the left panel we plot $T_{\rm SFR}$ vs. $M_\star$. The shaded horizontal band represents the age of the universe for the range of redshifts in our sample; if $T_{\rm SFR}$ is greater than the age of the universe at the redshift of the galaxy, then the galaxy cannot have formed all its stars at the current rate, and must have had a higher SFR in the past. Only objects with $M_\star \gtrsim 6 \times 10^{10}$ M$_\odot$ have $T_{\rm SFR}$ approximately the age of the universe. This upper limit on the time available for star formation suggests that most objects do not require declining star formation histories, though a CSF model may not be a reasonable fit for the most massive galaxies. These appear to be better described by a shallowly declining model, although the uncertainties in the SFR and stellar masses are large enough that this conclusion is not robust. Similar results are found from the SED modeling, as noted in §4.3; the issue is discussed in more detail by Shapley et al. (2005b). These results can be compared with those of a similar study involving galaxies at somewhat lower redshifts. Juneau et al. (2005) use galaxies from the Gemini Deep Deep Survey (GDDS) in the redshift range $0.8 < z < 2$ to compare $T_{\rm SFR}$ with the total time available for star formation, and find that galaxies with stellar masses of $\sim 6 - 30 \times 10^{10}$ M$_\odot$ must have had higher SFRs in the past at $z \lesssim 1.8$, while those with $\sim 2 - 6 \times 10^{10}$ M$_\odot$ reach a quiescent mode of star formation at $z \lesssim 1.1$. Thus galaxies of decreasing stellar mass stop forming stars in "burst" mode at later times; our results that most galaxies at $z \gtrsim 2$ are still in starburst mode are consistent with this broad conclusion. This consistency with the lower redshift sample is reassuring, since the clustering properties of the $z \sim 2$ galaxies indicate that they will become passively-evolving early-type galaxies like those found in the GDDS sample by $z \sim 1$ (Adelberger et al. 2005b).

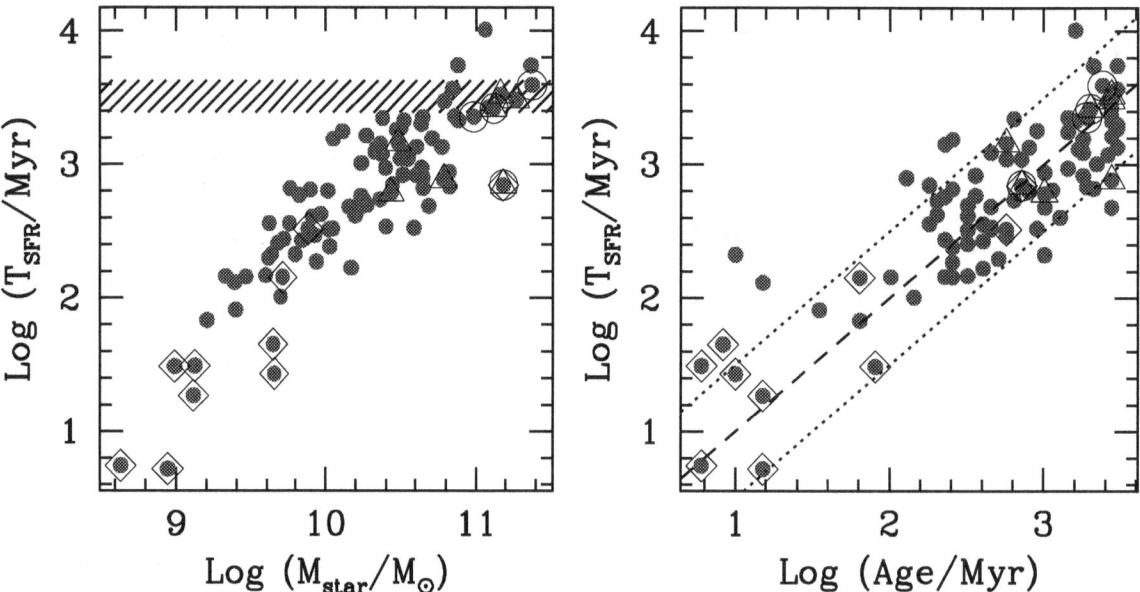

Figure 4.18 The star formation rate timescale $T_{\mathrm{SFR}} = M_\star/\mathrm{SFR}_{\mathrm{H}\alpha}$ inferred from the stellar masses and corrected Hα SFRs, plotted vs. stellar mass at left and vs. age at right. The shaded band in the left panel represents the age of the universe for the redshift range of the galaxies in the sample. The most massive galaxies have $T_{\mathrm{SFR}} \gtrsim t_{\mathrm{universe}}$, indicating that they may be better described by declining star formation histories. The plot at right provides a check of the consistency of the Hα SFRs and the primarily constant star formation models we use to fit the SEDs; for CSF models, T_{SFR} should be approximately equal to the age. The dashed line shows equal times, and the dotted lines on either side show the typical uncertainty in age. Symbols are as in Figure 4.5.

We can perhaps obtain additional constraints on the star formation histories by comparing T_{SFR} with the ages we obtain from the SED fitting. This is a consistency check for our SED fits and Hα SFRs, since most of the ages represent constant star formation models; if constant star formation at the current rate is indeed an adequate representation of the average star formation history, then T_{SFR} should be approximately equal to the age. We plot T_{SFR} vs. age in the right panel of Figure 4.18. The dashed line represents equal timescales; if objects fall significantly above this line, they cannot have formed all of their stars at their current rate over their inferred lifetime and must have had a past burst, while objects significantly below the line would have a current SFR higher than the past average. The dotted lines show the average uncertainty in the age, from our Monte Carlo simulations of the SED fits. Most of the objects fall between or near the dotted lines, suggesting that constant star formation over the age determined by the SED fit adequately describes the star formation histories of most of the galaxies in our sample, though the scatter is certainly large enough to allow for some declining star formation histories, as may be required for the most massive galaxies. It should also be noted that the tendency of a few of the youngest galaxies to fall above the dashed line is probably due to an underestimate of their ages, which cannot realistically be less than their dynamical times; for this set of objects, the average $t_{\mathrm{dyn}} \simeq 2r/\sigma = 36$ Myr (as compared to ~ 60 Myr for the entire sample).

The Hα equivalent width $W_{\mathrm{H}\alpha}$ provides an additional tool to investigate the star formation history. As the ratio of the Hα luminosity to the underlying stellar continuum, $W_{\mathrm{H}\alpha}$ is a measure of the ratio of the current to past average star formation. We determine $W_{\mathrm{H}\alpha}$ by taking the ratio of the Hα flux and our K-band continuum flux, after subtracting the contribution of Hα to the K-band magnitude. In calculating the equivalent widths we have applied the factor of 2 aperture correction to the Hα fluxes discussed above and in §4.2.1 (except in the cases of Q1623-BX455 and Q1623-BX502, for which twice the Hα flux slightly exceeds the K-band magnitude), but we have not applied an extinction correction; this is equivalent to the assumption that the nebular emission lines and the stellar continuum suffer the same attenuation. We plot $W_{\mathrm{H}\alpha}$ against our best-fit age from the SED fits in Figure 4.19. For constant star formation, $W_{\mathrm{H}\alpha}$ should decrease with age, as the stellar continuum increases while the Hα flux remains the same. There is considerable scatter in the $W_{\mathrm{H}\alpha}$–age comparison, but the probability that the data are uncorrelated is $P = 0.001$, for a significance of 3.3σ.

For simple star formation histories, the evolution of $W_{\mathrm{H}\alpha}$ with galaxy age can be predicted with models of stellar evolution and population synthesis. The solid black line in Figure 4.19 is the theoretically predicted dependence of $W_{\mathrm{H}\alpha}$ on age, from a Starburst99 (Leitherer et al. 1999) model with constant star formation, solar metallicity, and a Kroupa (2001) IMF, which gives very similar results to the Chabrier IMF we employ; the dashed blue line is the same, but for $Z = 0.4Z_\odot$. There is general agreement between the models and the data, but with a large amount of scatter. The equivalent width is a comparison of two quantities with very different timescales; the light from the stellar continuum gradually increases over time, while the Hα flux may vary stochastically on a much shorter timescale, in response to mergers, feedback, or accretion events. The scatter in the data with respect to the models is ~ 0.5 dex, which can be accounted for by a factor of ~ 2 change in the current star formation rate with respect to the past average (because a change in the Hα flux also affects the inferred continuum flux through the subtraction of Hα, the equivalent width can change by a larger factor than the star formation rate). A factor of ~ 2 is also the typical uncertainty in the star formation rate of individual objects.

The relative extinction of the nebular lines and stellar continuum probably also affects the results here. As mentioned above in the discussion of the star formation rates, we have not used the Calzetti et al. (2000) prescription of $E(B-V)_{\mathrm{stellar}} = 0.4E(B-V)_{\mathrm{neb}}$ for the extinction corrections because doing so results in a significant overestimate of the $\mathrm{SFR}_{\mathrm{H}\alpha}$ with respect to the SFRs from the UV continuum and our models (if we have overestimated the typical aperture correction, then there is room for additional nebular line extinction). Applying this additional extinction correction results in a typical increase of a factor of ~ 3 in $W_{\mathrm{H}\alpha}$; as can be seen in Figure 4.19, the mean value of $W_{\mathrm{H}\alpha}$ is somewhat below the CSF predictions at

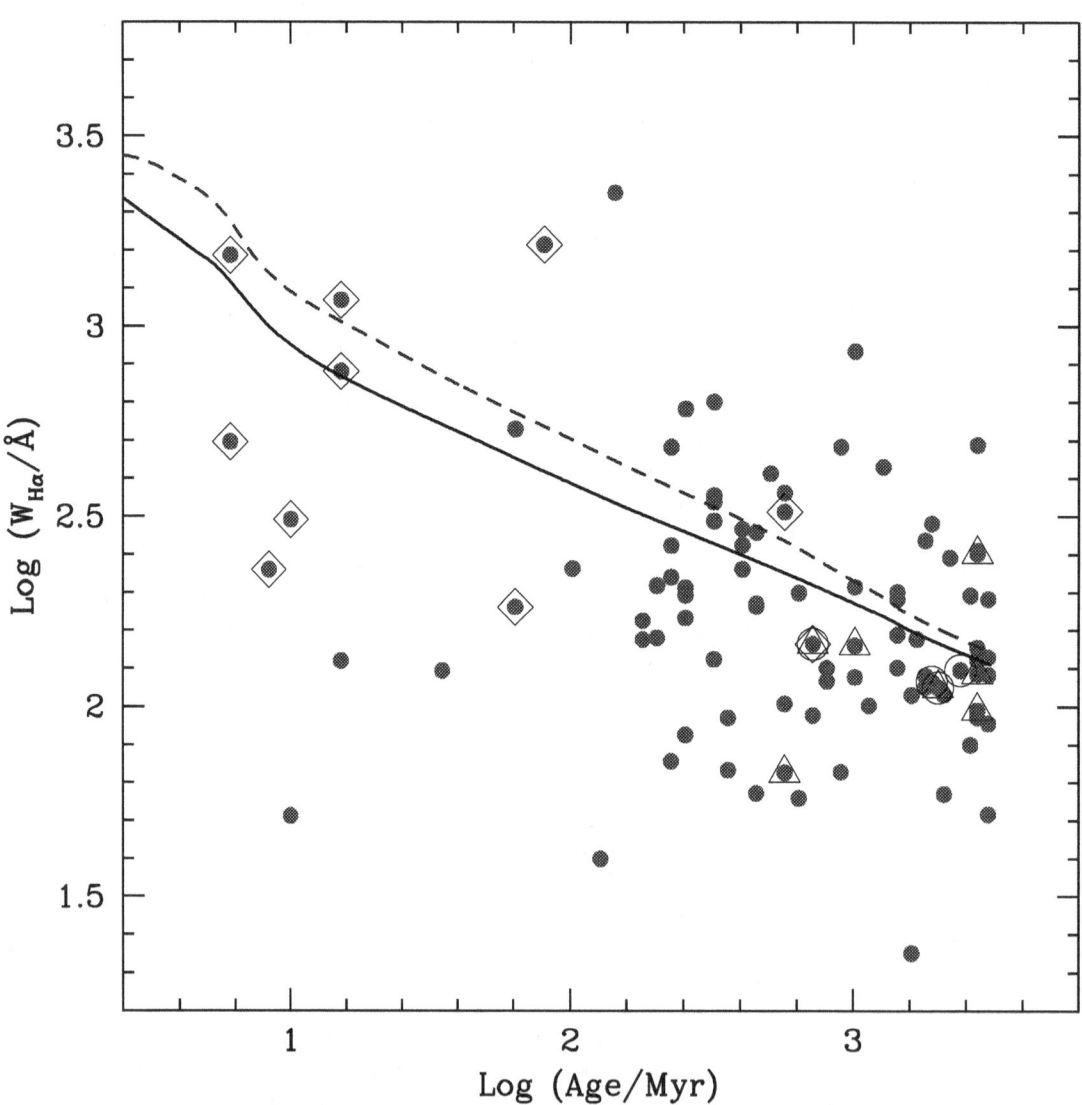

Figure 4.19 A comparison of Hα equivalent width and age from the SED modeling. The lines show the predicted $W_{H\alpha}$ as a function of age for constant star formation, from Starburst99 models with solar (solid black line) and 0.4 solar (dashed blue line) metallicity. The large scatter of the data with respect to the models is probably caused by variations in the SFR as well as observational uncertainties. Differential extinction to H II regions is likely to be a factor as well. Symbols are as in Figure 4.5.

a given age, but not usually by a factor of 3. It is possible that the H II regions do suffer some smaller amount of additional extinction, however, and this may explain the larger number of objects in our sample that fall below the predictions. This should be more significant for the older objects, as the stars in young galaxies have not had as much time to migrate away from the dusty regions in which they form. It appears from Figure 4.19, however, that it is the youngest objects that have systematically lower equivalent widths than predicted by the models. These are also the objects for which $W_{\mathrm{H}\alpha}$ is the most uncertain, however. The typical uncertainty in $W_{\mathrm{H}\alpha}$ is $\sim 40\%$, but this approaches $\sim 100\%$ for the galaxies in which the Hα flux makes up most of the K-band light; comparisons of ages and equivalent widths should be regarded as highly uncertain in this regime. We also note that the most anomalous point, in the lower left corner, corresponds to Q1700-BX681, which is not well-fit by any model SED and therefore has a very uncertain age. As mentioned above, we may have somewhat underestimated the ages of the youngest objects in general, as the ages cannot be less than the dynamical timescale $t_{\mathrm{dyn}} \sim 35$ Myr.

Nothing in our results contradicts the theory that the current star formation rate is generally representative of the past average for most of our sample, although stochastic variations are likely. We are not able to strongly discriminate between star formation histories, however; a shallowly declining star formation history would also result in equivalent widths somewhat below the CSF predictions, and this may also be a factor for some of the objects in our sample, particularly the oldest and most massive. Extrapolating forward in time, the star formation rates of the galaxies in our sample will certainly decline as they lose their gas to star formation or winds, and eventually become the early-type galaxies of today (Adelberger et al. 2005b).

4.5.2 Gas and Baryonic Masses

In star-forming galaxies in the local universe, the surface densities of star formation and gas are observed to follow a Schmidt (1959) law, $\Sigma_{\mathrm{SFR}} = A\Sigma_{\mathrm{gas}}^{N}$, over more than 6 orders of magnitude in Σ_{SFR} (Kennicutt 1998b). This empirical relation is usually explained by a model in which the SFR scales with density-dependent gravitational instabilities in the gas. The correlation has not yet been tested at high redshift because of the lack of measurements of gas masses, although the one well-studied example, the lensed $z = 2.7$ LBG MS1512-cB58, appears to be consistent with the local Schmidt law (Baker et al. 2004). Assuming that the galaxies in our sample obey such a law, we can use the SFRs and the galaxy sizes $r_{\mathrm{H}\alpha}$ measured from the spatial extent of the Hα emission to compute their star formation densities, and thus estimate their gas densities and masses. We use the global Schmidt law of Kennicutt (1998b):

$$\Sigma_{\mathrm{SFR}} = 2.5 \times 10^{-4} \left(\frac{\Sigma_{\mathrm{gas}}}{1\ \mathrm{M}_\odot\ \mathrm{pc}^{-2}} \right)^{1.4} \mathrm{M}_\odot\ \mathrm{yr}^{-1}\ \mathrm{kpc}^{-2} \qquad (4.2)$$

in combination with the conversion from Hα luminosity to SFR from the same paper,

$$\mathrm{SFR}\ (\mathrm{M}_\odot\ \mathrm{yr}^{-1}) = \frac{\mathrm{L}(\mathrm{H}\alpha)}{1.26 \times 10^{41}\ \mathrm{erg}\ \mathrm{s}^{-1}} \qquad (4.3)$$

to create an IMF-independent relation between our observed Hα luminosity per unit area and the gas surface density

$$\Sigma_{\mathrm{gas}} = 1.6 \times 10^{-27} \left(\frac{\Sigma_{\mathrm{H}\alpha}}{\mathrm{erg}\ \mathrm{s}^{-1}\ \mathrm{kpc}^{-2}} \right)^{0.71} \mathrm{M}_\odot\ \mathrm{pc}^{-2}. \qquad (4.4)$$

The radii used by Kennicutt (1998b) to compute surface densities approximately coincide with the edge of the galaxies' Hα-emitting disks; for our surface densities, we take an area equal to the square of the FWHM of the Hα emission (note that this $r_{\mathrm{H}\alpha}$ is a factor of 2.4 larger than the radius used for the calculation of dynamical masses, for which we used the rms of the Hα emission to correspond with σ, the rms in velocity). We then take $M_{\mathrm{gas}} = \Sigma_{\mathrm{gas}} r_{\mathrm{H}\alpha}^2$ as an estimate of the gas mass associated with star formation, and calculate the gas fraction $\mu \equiv M_{\mathrm{gas}}/(M_{\mathrm{gas}} + M_\star)$. The removal of the IMF dependence from the Schmidt law (which assumes a Salpeter IMF in the conversion from Hα luminosity to SFR) facilitates comparison between gas, dynamical, and stellar masses, as we discuss below. Because the sizes and the corrected Hα fluxes of the objects are uncertain by up to a factor of ~ 2, the gas masses of individual objects are uncertain by a factor of ~ 3, and the typical fractional uncertainty in μ is $\sim 50\%$. Systematic uncertainties due to the scatter in the Schmidt law itself (0.3 dex) are an additional source of error. We focus here on overall trends, which depend on large numbers of objects and so are better determined than the parameters of individual galaxies; for example, the error in the mean gas fraction of a subsample of ~ 15 objects is 15% or less.

Although the above equations omit the step of calculating the SFR surface density for our sample, it is worthwhile to do so for comparison with local galaxies. After converting our SFRs to a Salpeter IMF by multiplying by 1.8, we find a mean $\langle \Sigma_{\mathrm{SFR}} \rangle = 2.9$ M$_\odot$ yr^{-1} kpc^{-2}. As shown in Figure 4.20, our observed distribution is similar to the sample of local starburst galaxies studied by Kennicutt (1998b), with the exception that we do not see objects with $\Sigma_{\mathrm{SFR}} \gtrsim 20$ M$_\odot$ yr^{-1} kpc^{-2}; the upper cutoff of our distribution is an order of magnitude lower than is seen locally. This cutoff is easily understood. The nearby galaxies with the highest values of Σ_{SFR} are the luminous and ultra-luminous IR galaxies (LIRGS and ULIRGs), which have bolometric luminosities $\sim 10^{11}$–10^{12} L$_\odot$. The high redshift counterparts of such objects are the submillimeter-detected SCUBA galaxies (e.g., Chapman et al. 2005); although two or three such objects may be present in the current sample,[2] we would not infer their extreme SFRs from their UV and Hα properties alone (Swinbank et al. 2004; Chapman et al. 2005; Reddy et al. 2005). Our inability to resolve star formation on small scales also limits our ability to detect objects with high Σ_{SFR}, although because most of our objects are resolved, and the smallest objects have lower than average SFRs, this is unlikely to be a major factor. We also note that all of our objects have $\Sigma_{\mathrm{SFR}} > 0.1$ M$_\odot$ yr^{-1} kpc^{-2}; starburst-driven superwinds are observed to be ubiquitous in galaxies with SFR densities above this threshold (Heckman 2002).

The mean gas mass is $\langle M_{\mathrm{gas}} \rangle = 2.1 \times 10^{10}$ M$_\odot$, slightly lower than the mean stellar mass of $\langle M_\star \rangle = 3.6 \times 10^{10}$ M$_\odot$. The distributions of the two masses are shown in the histograms in the upper left panel of Figure 4.21, where the narrower blue histogram shows the gas masses and the broader red histogram shows the stellar masses. The range of gas masses is clearly smaller than the range of stellar masses; this is because the dispersions in star formation rate and size are smaller than the more than 2 orders of magnitude variation we see in stellar mass. The two masses are plotted against each other in the upper right panel of Figure 4.21, where it is apparent that the $M_{\mathrm{dyn}}/M_\star > 7$ objects (open diamonds) have large gas masses, and are exceptions to a general trend of increasing gas mass with increasing stellar mass. As discussed above, the absence of points in the lower left corner of this plot is probably a selection effect, as low-mass galaxies with low star formation rates are unlikely to be detected in our K-band images (they may fall below our $\mathcal{R} < 25.5$ magnitude limit for selection as well). We next plot the gas fraction μ vs. stellar mass in the lower left panel of Figure 4.21 and μ vs. age in the lower right. The trends of decreasing gas fraction with increasing stellar mass and age are strong, supporting our hypothesis that the $M_{\mathrm{dyn}} \gg M_\star$ objects are young starbursts with high gas fractions (low-mass galaxies with low gas fractions may exist, but are probably too faint to be detected by our survey). Local galaxies with low stellar masses are also observed to have higher gas fractions (McGaugh & de Blok 1997; Bell & de Jong 2000).

Finally, we compare the estimated baryonic masses $M_{\mathrm{bar}} = M_{\mathrm{gas}} + M_\star$ with the dynamical masses in Figure 4.22, to see if the addition of the gas mass improves the agreement. We use the same axes as

---

[2]Approximately 2% of UV-selected galaxies are confirmed submillimeter sources, while $\sim 50\%$ of submillimeter galaxies are selected by the UV criteria (Reddy et al. 2005; Chapman et al. 2005).

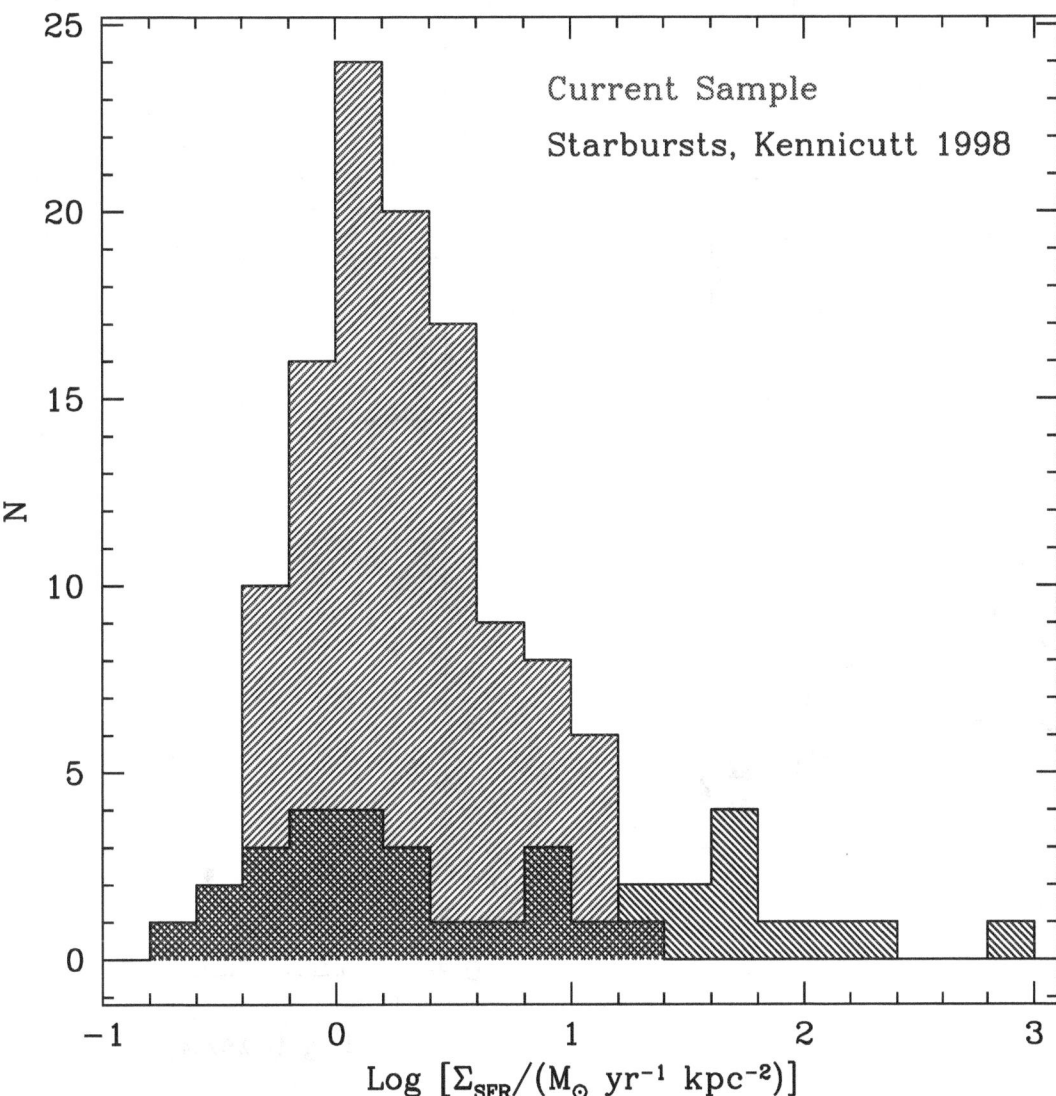

Figure 4.20 A comparison of the star formation surface densities Σ_{SFR} of our sample and the starbursts of Kennicutt (1998b). We use the Hα luminosity to determine SFRs, whereas the SFRs of the local sample are determined primarily from their FIR emission. This accounts for the lack of objects with the highest values of Σ_{SFR} in our sample, since Hα underestimates the SFR of luminous IR galaxies even after correcting for extinction. We use a Salpeter IMF for consistency with the low redshift sample.

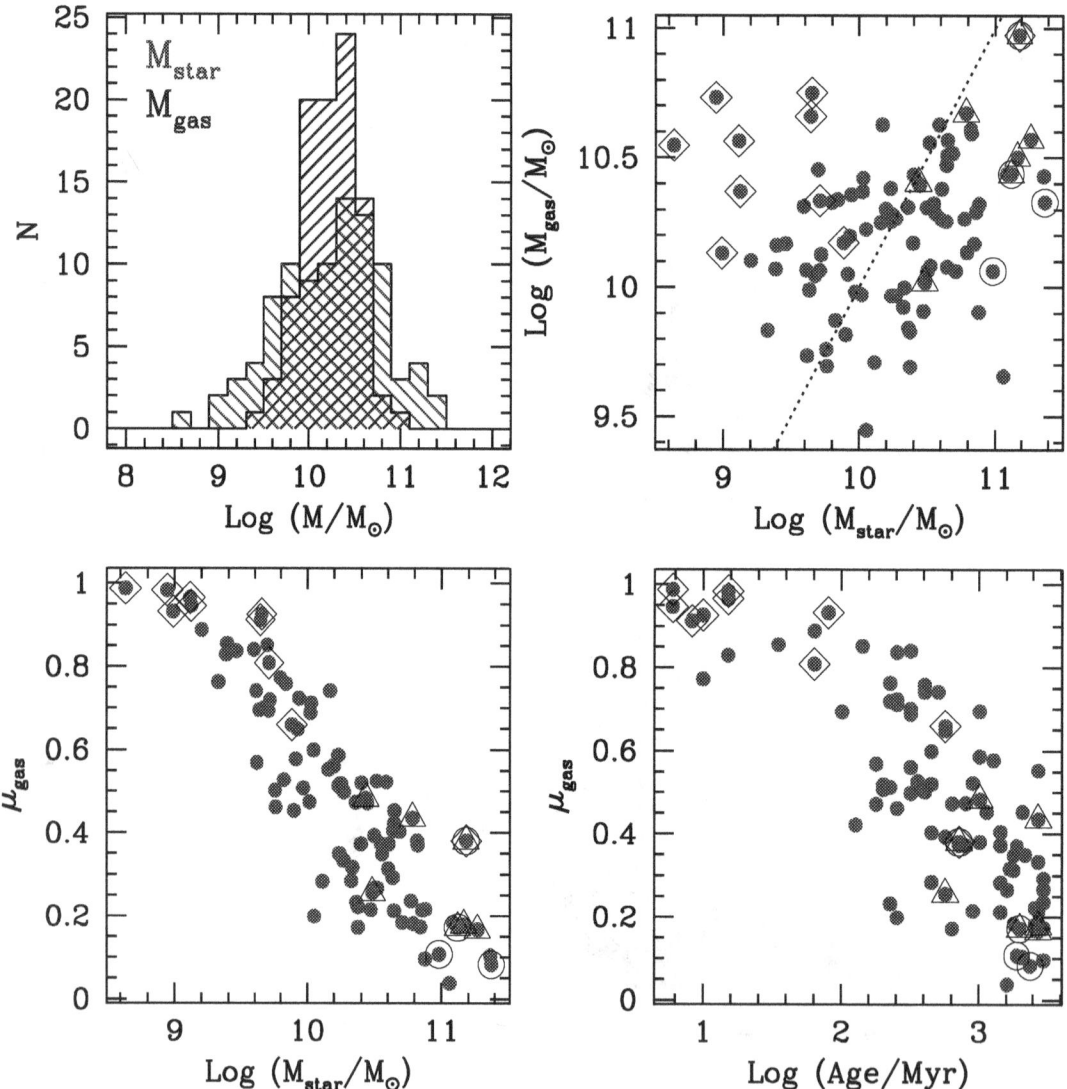

Figure 4.21 The relation of the gas mass inferred from the Schmidt law to the stellar population parameters. Upper left: Histograms of the gas (blue) and stellar (red) masses, showing that the range in stellar mass is significantly broader than the range in stellar mass. Upper right: Gas mass vs. stellar mass. Gas masses are relatively constant across the sample, but when the galaxies with $M_{\rm dyn}/M_\star > 7$ are not considered we see an increase in gas mass with stellar mass, because of the increase in SFR with stellar mass. Lower panels: The gas fraction strongly decreases with increasing stellar mass (left) and age (right). Symbols are as in Figure 4.5.

Figure 4.5 to facilitate comparison, and again mark the line of equal masses (dashed line) and a factor of 7 difference between the two masses (dotted lines). The agreement between the masses is greatly improved with the addition of the gas; there is a significant correlation (4σ, with probability $P = 8 \times 10^{-5}$ that the data are uncorrelated), and the masses are within a factor of 3 for 80% of the sample (we do not include the objects with limits on σ, though most of these appear to be consistent with the rest of the data, and those that are not are highly uncertain). There are only two objects for which $M_{\mathrm{dyn}}/M_{\mathrm{bar}} > 7$; these are the AGN Q1700-MD94 (upper right), and the galaxy Q1623-BX376, which has an anomalously large velocity dispersion that may be influenced by its complicated spatial structure (§4.4.1, Erb et al. 2003). For most objects, the dynamical masses are somewhat lower than the baryonic masses. After excluding AGN (open circles), we find $\langle M_{\mathrm{gas}} + M_{\star} \rangle = 6.1 \times 10^{10}$ $M_{\odot}$, while the mean dynamical mass $\langle M_{\mathrm{dyn}} \rangle = 4.5 \times 10^{10}$ $M_{\odot}$ is 1.7 times lower. This difference may indicate that too small a size has been used to determine the dynamical masses; after correction for the seeing, the typical isophotal radius used for the K-band photometry is ~ 2.5 times the rms $r_{\mathrm{H}\alpha} = FWHM/2.4$ we use for the dynamical masses. Such a larger size may be more appropriate for comparisons with properties determined from global star formation rates. Another possibility is that the constant we use in Eq. 4.1 is too small. Alternatively or additionally, this may simply be an indication that the velocity dispersions do not trace the full potential, as we have mentioned several times previously. The gas masses are almost certainly underestimates as well. The gas masses derived from the Schmidt law include only the cold gas associated with current star formation; in a typical local disk galaxy, $\sim 40\%$ of the total gas mass is not included by the Schmidt law (Martin & Kennicutt 2001), and this fraction could plausibly be higher in the young starbursts in our sample, which may still be experiencing significant infall of cooling gas from the halo. A simple argument from clustering analysis is consistent with this interpretation of our results. Clustering properties have shown that the typical halo mass of the $z \sim 2$ galaxies is $\sim 2 \times 10^{12}$ $M_{\odot}$ (Adelberger et al. 2005b), and the baryon to dark matter ratio (0.17, Spergel et al. 2003) therefore implies a typical baryonic mass of $\sim 3.4 \times 10^{11}$ $M_{\odot}$, ~ 6 times higher than our average $M_{\mathrm{gas}} + M_{\star}$. Substantial additional gas mass (a typical increase of a factor of ~ 12, since $M_{\mathrm{gas}} \sim M_{\star}$ on average) is required to bring the average baryonic mass into agreement with the clustering predictions. Even in most local galaxies, however, not all of the baryons predicted by the baryon to dark matter ratio are accounted for; the Milky Way, for example, has a baryon to total mass ratio of 3–4% in a $\sim 10^{12}$ $M_{\odot}$ halo (Mo & Mao 2004).

4.6 Summary and Discussion

We have presented the analysis of near-IR Hα spectra of 114 star-forming galaxies at $z \sim 2$, and looked for trends between properties revealed by Hα and the results of the fitting of model SEDs to multiwavelength photometry of 93 of the 114 objects, from which we estimate stellar masses, ages, extinction and star formation rates. The sample covers nearly three orders of magnitude in stellar mass, from $M_{\star} = 4 \times 10^{8}$ $M_{\odot}$ to $M_{\star} = 2 \times 10^{11}$ $M_{\odot}$, and best-fit ages range from a few Myr to the age of the universe at the redshift of the sample, ~ 2.8 Gyr. Our main conclusions are as follows:

1. The sample has a mean Hα velocity dispersion $\langle \sigma \rangle = 112$ km s^{-1}. From σ and the spatial extent of the Hα emission we estimate dynamical masses, finding $\langle M_{\mathrm{dyn}} \rangle = 4 \times 10^{10}$ $M_{\odot}$, excluding AGN. Stellar and dynamical masses agree to within a factor of 7 for most objects, consistent with observational and systematic uncertainties. The masses are correlated with 3σ significance. However, 15% of the galaxies have $M_{\mathrm{dyn}} > 7 \times M_{\star}$; these objects have low stellar masses, young ages, and tend to have high Hα equivalent widths, suggesting that they are young galaxies with large gas fractions.

2. We combine our Hα results with LBGs at $z \sim 3$ and other samples from the literature, and find that rest-frame optical luminosity (corrected for extinction) and velocity dispersion are correlated with 4σ significance. The scatter is larger than can be accounted for by observational errors and prevents a

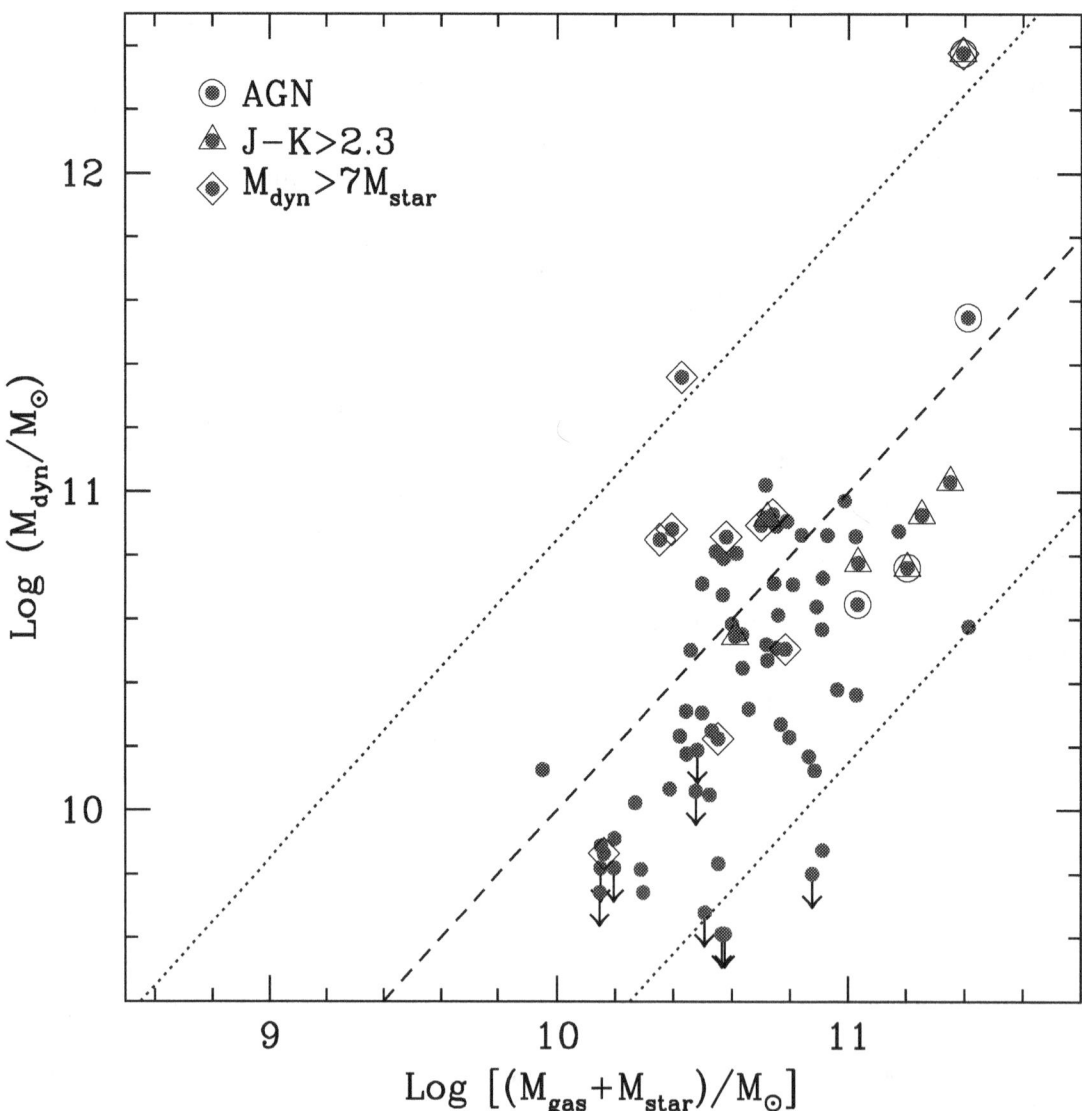

Figure 4.22 Baryonic mass $M_{\mathrm{gas}} + M_\star$ vs. M_{dyn}; compare Figure 4.5. The dashed line shows equal masses, and the dotted lines show a factor of 7 difference between the masses. The correlation is significant at the $4\,\sigma$ level, and the masses of 80% of the objects agree to within a factor of 3. The average dynamical mass is slightly lower than the average baryonic mass, suggesting that the velocity dispersion does not trace the full distribution of baryonic mass, that the size we use to determine dynamical masses is too small, or that we are using an incorrect average value of the constant in the dynamical mass formula. Symbols are as in Figure 4.5.

robust determination of the slope, though it is not inconsistent with the local relation $L \propto \sigma^4$. The high redshift galaxies have lower velocity dispersions at a given luminosity, or much higher luminosities at a given velocity dispersion, than the local Faber-Jackson relation.

3. The sample has a mean star formation rate from extinction-corrected Hα luminosity $\langle \text{SFR}_{\text{H}\alpha} \rangle = 31$ M$_\odot$ yr^{-1}. SFR increases with increasing stellar mass and at brighter K magnitudes, to $\langle \text{SFR}_{\text{H}\alpha} \rangle \sim 65$ M$_\odot$ yr^{-1} for galaxies with $K_s < 20$. SFRs determined from Hα, the rest-frame UV continuum and X-rays are strongly correlated, and give the same average value.

4. Using the empirical correlation between star formation rate per unit area and gas surface density, we estimate gas masses and gas fractions for the galaxies in the sample. The gas fraction strongly increases with decreasing stellar mass, and the mass in gas and stars is significantly better correlated with the dynamical mass than is the stellar mass alone.

4. Fourteen of the 114 galaxies with Hα spectra, or 12%, have spatially resolved and tilted emission lines. On average, the observed amplitude of the velocity shear v_c is approximately equal to the velocity dispersion σ. If the galaxies are rotating disks and follow the local relations between the true circular velocity V_c and σ, we underestimate V_c by an average factor of ~ 2, and the galaxies have $\langle V_c \rangle \sim 190$ km s^{-1}. However, merging may contribute significantly to the observed shear.

5. The rest-frame UV interstellar absorption lines are blueshifted with respect to Hα by an average of ~ 150 km s^{-1}, and the Lyα emission line, when present, is redshifted by an average of ~ 470 km s^{-1}. These offsets indicate the presence of galactic-scale outflows. Outflow velocity is not significantly correlated with stellar or dynamical mass, velocity dispersion, star formation rate, or star formation rate per unit area. However, galaxies with velocity shear have interstellar lines that are less blueshifted with respect to Hα than average, and a larger than expected fraction of galaxies with $v_{\text{abs}} \sim 0$ have shear.

The picture that emerges from the combination of Hα spectra and SED modeling is one of galaxies with a broad range in stellar mass, while kinematic measurements and estimates of gas fractions imply that much of the variation is due to age and the amount of gas converted into stars rather than to real differences in mass across the sample. The stellar masses of the galaxies considered here vary by up to a factor of 500, while the baryonic and dynamical masses vary by a factor of ~ 30. The halo masses inferred from the clustering properties of the $z \sim 2$ galaxies also support a relatively narrow range in total mass, with halos of $10^{11.8} - 10^{12.2}$ M$_\odot$ spanning only a factor of 2.5 (Adelberger et al. 2005b).

The star formation rates per unit area are uniformly above the threshold needed to support galactic outflows, in agreement with the kinematic signatures of outflows in nearly all the galaxies in the sample. Evidence of outflows is observed across the full age range of the sample, indicating that either the outflows can be sustained by relatively constant star formation over a timescale of up to ~ 3 Gyr, or that they come and go as variations in the star formation rate allow. In this case the galaxies would drop out of our sample during their quiescent phases. The duty cycle of such wind-producing star formation can be estimated by the relative fractions of star-forming and passive galaxies at $z \sim 2$. Reddy et al. (2005) have recently compared samples of galaxies selected by different criteria at $z \sim 2$; their results suggest that less than 30% of galaxies to $K < 21$ are passive, though the fraction may be $\sim 50\%$ for the most massive, K-bright objects. In either case, the winds must be sustained over a substantial fraction of the lifetime of the oldest galaxies in the sample; some balance between the outflows and star formation is clearly required, to maintain sufficiently high SFRs without driving all of the gas out of the galaxy. Winds maintained over $\sim$ Gyr timescales are also likely to have a profound impact on the IGM, depending on how far they penetrate into the IGM and on the fraction of metals that escape the galaxy.

Because of uncertainties in the mass and velocity distributions of the $z \sim 2$ galaxies, our kinematic measurements, and the dynamical masses in particular, are probably the most in need of improvement among our results. Such improvements will require deep observations of a large sample of objects, and in most cases high angular resolution will be required in order to discriminate between plausible models of the velocity field. Near-IR multiobject and integral field spectrographs, occasionally with adaptive optics, will make such observations feasible, and we anticipate that the galaxies presented here will be a useful sample from which to select targets for such observations. These kinematic measurements will provide otherwise unobtainable insights about the prevalence of mergers at high redshift and the growth of massive disk galaxies, and when combined with multiwavelength studies which will improve our understanding of star formation and dust, will lead to an increased understanding of the assembly of stellar mass at high redshift.

We thank Andrew Blain, Jonathan Bird, David Kaplan, and Shri Kulkarni for obtaining near-IR images of some of our targets, and the staffs of the Keck and Palomar observatories for their assistance with the observations. CCS, DKE and NAR have been supported by grant AST03-07263 from the U.S. National Science Foundation and by the David and Lucile Packard Foundation. AES acknowledges support from the Miller Institute for Basic Research in Science, and KLA from the Carnegie Institution of Washington.

Table 4.1. Galaxies Observed

Object	R.A.	Dec.	$z_{H\alpha}$	$\mathcal{R}^a$	$G - \mathcal{R}^a$	$U_n - G^a$	$K_s^{\,b}$	$J - K_s^{\,b}$	Exposure time (s)
CDFb-BN88[c]	00:53:52.87	12:23:51.25	2.2615	23.14	0.29	0.68	...	...	12 × 720
HDF-BX1055	12:35:59.59	62:13:07.50	2.4899	24.09	0.24	0.81	...	...	2 × 900
HDF-BX1084	12:36:13.57	62:12:21.48	2.4403	23.24	0.26	0.72	...	...	5 × 900
HDF-BX1085	12:36:13.33	62:12:16.31	2.2407	24.50	0.33	0.87	...	...	5 × 900
HDF-BX1086	12:36:13.41	62:12:18.84	2.4435	24.64	0.41	1.09	...	...	5 × 900
HDF-BX1277	12:37:18.59	62:09:55.54	2.2713	23.87	0.14	0.61	21.26	0.73	3 × 900
HDF-BX1303	12:37:11.20	62:11:18.67	2.3003	24.72	0.11	0.81	21.03	2.31	2 × 900
HDF-BX1311	12:36:30.54	62:16:26.12	2.4843	23.29	0.21	0.81	20.48	1.56	4 × 900
HDF-BX1322	12:37:06.54	62:12:24.94	2.4443	23.72	0.31	0.57	20.95	2.16	6 × 900
HDF-BX1332	12:37:17.13	62:11:39.95	2.2136	23.64	0.32	0.92	20.68	1.77	3 × 900
HDF-BX1368	12:36:48.24	62:15:56.24	2.4407	23.79	0.30	0.96	20.63	1.81	4 × 900
HDF-BX1376	12:36:52.96	62:15:45.55	2.4294	24.48	0.01	0.70	22.13	1.02	4 × 900
HDF-BX1388	12:36:44.84	62:17:15.84	2.0317	24.55	0.27	0.99	19.95	1.78	2 × 900
HDF-BX1397	12:37:04.12	62:15:09.84	2.1328	24.12	0.14	0.76	20.87	1.08	3 × 900
HDF-BX1409	12:36:47.41	62:17:28.70	2.2452	24.66	0.49	1.17	20.07	2.31	2 × 900
HDF-BX1439	12:36:53.66	62:17:24.27	2.1865	23.90	0.26	0.79	19.72	2.16	4 × 900
HDF-BX1479	12:37:15.42	62:16:03.88	2.3745	24.39	0.16	0.79	21.30	1.57	5 × 900
HDF-BX1564	12:37:23.47	62:17:20.02	2.2225	23.28	0.27	1.01	19.62	1.69	2 × 900
HDF-BX1567	12:37:23.17	62:17:23.89	2.2256	23.50	0.18	1.05	20.18	1.23	2 × 900
HDF-BX305	12:36:37.13	62:16:28.36	2.4839	24.28	0.79	1.30	20.14	2.59	4 × 900
HDF-BMZ1156	12:37:04.34	62:14:46.28	2.2151	24.62	-0.01	-0.21	20.33	1.71	16 × 720
Q0201-B13[c]	02:03:49.25	11:36:10.58	2.1663	23.34	0.02	0.69	...	...	2 × 900
Q1307-BM1163	13:08:18.04	29:23:19.34	1.4105	21.66	0.20	0.35	...	...	2 × 900
Q1623-BX151	16:25:29.61	26:53:45.01	2.4393	24.60	0.14	0.88	...	...	4 × 900
Q1623-BX214	16:25:33.67	26:53:53.52	2.4700	24.06	0.39	1.12	...	...	4 × 900
Q1623-BX215	16:25:33.80	26:53:50.66	2.1814	24.45	0.26	0.53	...	...	4 × 900
Q1623-BX252	16:25:36.96	26:45:54.86	2.3367	25.06	0.07	0.55	...	...	3 × 900
Q1623-BX274	16:25:38.20	26:45:57.14	2.4100	23.23	0.25	0.89	...	...	3 × 900
Q1623-BX344	16:25:43.93	26:43:41.98	2.4224	24.42	0.39	1.25	...	...	2 × 900
Q1623-BX366	16:25:45.09	26:43:46.95	2.4204	23.84	0.41	1.03	...	...	2 × 900
Q1623-BX376	16:25:45.59	26:46:49.26	2.4085	23.31	0.24	0.75	20.84	1.61	4 × 900
Q1623-BX428	16:25:48.41	26:47:40.20	2.0538	23.95	0.13	1.06	20.72	1.02	4 × 900
Q1623-BX429	16:25:48.65	26:45:14.47	2.0160	23.63	0.12	0.56	20.94	1.46	2 × 900
Q1623-BX432	16:25:48.73	26:46:47.28	2.1817	24.58	0.10	0.53	21.48	1.76	4 × 900
Q1623-BX447	16:25:50.37	26:47:14.28	2.1481	24.48	0.17	1.14	20.55	1.66	3 × 900
Q1623-BX449	16:25:50.53	26:46:59.97	2.4188	24.86	0.20	0.61	21.35	1.88	3 × 900
Q1623-BX452	16:25:51.00	26:44:20.00	2.0595	24.73	0.20	0.90	20.56	2.05	3 × 900
Q1623-BX453	16:25:50.84	26:49:31.40	2.1816	23.38	0.48	0.99	19.76	1.65	3 × 900

Table 4.1—Continued

Object	R.A.	Dec.	$z_{H\alpha}$	$\mathcal{R}^{a}$	$G - \mathcal{R}^{a}$	$U_n - G^{a}$	K_s^{b}	$J - K_s^{b}$	Exposure time (s)
Q1623-BX455	16:25:51.66	26:46:54.88	2.4074	24.80	0.35	0.87	21.56	1.83	2 × 900
Q1623-BX458	16:25:51.58	26:46:21.39	2.4194	23.41	0.28	0.85	20.52	1.23	4 × 900
Q1623-BX472	16:25:52.87	26:46:39.63	2.1142	24.58	0.16	0.91	20.80	1.91	4 × 900
Q1623-BX502	16:25:54.38	26:44:09.25	2.1558	24.35	0.22	0.50	22.04	0.99	3 × 900
Q1623-BX511	16:25:56.11	26:44:44.57	2.2421	25.37	0.42	1.05	21.78	...	4 × 900
Q1623-BX513	16:25:55.86	26:46:50.30	2.2473	23.25	0.26	0.68	20.21	1.83	2 × 900
Q1623-BX516	16:25:56.27	26:44:08.19	2.4236	23.94	0.30	0.82	20.41	2.29	3 × 900
Q1623-BX522	16:25:55.76	26:44:53.28	2.4757	24.50	0.31	1.19	20.75	2.03	4 × 900
Q1623-BX528	16:25:56.44	26:50:15.44	2.2682	23.56	0.25	0.71	19.75	1.79	4 × 900
Q1623-BX543	16:25:57.70	26:50:08.59	2.5211	23.11	0.44	0.96	20.54	1.31	4 × 900
Q1623-BX586	16:26:01.52	26:45:41.58	2.1045	24.58	0.32	0.87	20.84	1.79	4 × 900
Q1623-BX599	16:26:02.54	26:45:31.90	2.3304	23.44	0.22	0.80	19.93	2.09	4 × 900
Q1623-BX663	16:26:04.58	26:48:00.20	2.5373	24.14	0.24	1.02	19.92	2.59	3 × 900
Q1623-MD107	16:25:53.87	26:45:15.46	2.5373	25.35	0.12	1.43	22.43	1.72	4 × 900
Q1623-MD66	16:25:40.39	26:50:08.88	2.1075	23.95	0.37	1.40	20.15	1.74	3 × 900
Q1700-BX490	17:01:14.83	64:09:51.69	2.3960	22.88	0.36	0.92	19.99	1.54	3 × 900
Q1700-BX505	17:00:48.22	64:10:05.86	2.3089	25.17	0.45	1.28	20.85	2.17	4 × 900
Q1700-BX523	17:00:41.71	64:10:14.88	2.4756	24.51	0.46	1.28	20.93	1.86	4 × 900
Q1700-BX530	17:00:36.86	64:10:17.38	1.9429	23.05	0.21	0.69	19.92	1.23	4 × 900
Q1700-BX536	17:01:08.94	64:10:24.95	1.9780	23.00	0.21	0.79	19.71	1.31	4 × 900
Q1700-BX561	17:01:04.18	64:10:43.83	2.4332	24.65	0.19	1.04	19.87	2.43	2 × 900
Q1700-BX581	17:01:02.73	64:10:51.30	2.4022	23.87	0.28	0.62	20.79	1.94	2 × 900
Q1700-BX681	17:01:33.76	64:12:04.28	1.7396	22.04	0.19	0.40	19.18	1.27	4 × 900
Q1700-BX691	17:01:06.00	64:12:10.27	2.1895	25.33	0.22	0.66	20.68	1.82	4 × 900
Q1700-BX717	17:00:56.99	64:12:23.76	2.4353	24.78	0.20	0.61	21.89	1.49	4 × 900
Q1700-BX759	17:00:59.55	64:12:55.45	2.4213	24.43	0.36	1.29	21.23	1.35	2 × 900
Q1700-BX794	17:00:47.30	64:13:18.70	2.2473	23.60	0.35	0.58	20.53	1.37	3 × 900
Q1700-BX917	17:01:16.11	64:14:19.80	2.3069	24.43	0.28	0.95	20.03	1.95	3 × 900
Q1700-MD103	17:01:00.21	64:11:55.58	2.3148	24.23	0.46	1.49	19.94	1.96	900 + 600
Q1700-MD109	17:01:04.48	64:12:09.29	2.2942	25.46	0.26	1.44	21.77	1.75	4 × 900
Q1700-MD154	17:01:38.39	64:14:57.37	2.6291	23.23	0.73	1.91	19.68	2.08	3 × 900
Q1700-MD174	17:00:54.54	64:16:24.76	2.3423	24.56	0.32	1.50	19.90	...	4 × 900
Q1700-MD69	17:00:47.62	64:09:44.78	2.2883	24.85	0.37	1.50	20.05	2.60	4 × 900
Q1700-MD94	17:00:42.02	64:11:24.22	2.3362	24.72	0.94	2.06	19.65	2.46	3 × 900
Q2343-BM133	23:46:16.18	12:48:09.31	1.4774	22.59	0.00	0.19	20.50	0.66	3 × 900
Q2343-BM181	23:46:27.03	12:49:19.65	1.4951	24.77	0.12	0.29	...	...	4 × 900
Q2343-BX163	23:46:04.78	12:45:37.78	2.1213	24.07	-0.01	0.71	21.38	0.97	4 × 900
Q2343-BX169	23:46:05.03	12:45:40.77	2.2094	23.11	0.19	0.72	20.75	1.05	4 × 900

Table 4.1—Continued

Object	R.A.	Dec.	$z_{H\alpha}$	$\mathcal{R}^a$	$G - \mathcal{R}^a$	$U_n - G^a$	$K_s{}^b$	$J - K_s{}^b$	Exposure time (s)
Q2343-BX182	23:46:18.04	12:45:51.11	2.2879	23.74	0.14	0.56	21.60	0.92	4 × 900
Q2343-BX236	23:46:18.71	12:46:15.97	2.4348	24.28	0.14	0.71	21.25	1.60	3 × 900
Q2343-BX336	23:46:29.53	12:47:04.76	2.5439	23.91	0.40	1.15	20.80	1.75	4 × 900
Q2343-BX341	23:46:23.24	12:47:07.97	2.5749	24.21	0.38	0.89	21.40	2.07	3 × 900
Q2343-BX378	23:46:33.90	12:47:26.20	2.0441	24.80	0.26	0.55	21.90	1.30	4 × 900
Q2343-BX389	23:46:28.90	12:47:33.55	2.1716	24.85	0.28	1.26	20.18	2.74	3 × 900
Q2343-BX390	23:46:24.72	12:47:33.80	2.2313	24.36	0.24	0.79	21.29	1.73	4 × 900
Q2343-BX391	23:46:28.07	12:47:31.82	2.1740	24.51	0.25	1.01	21.93	1.35	3 × 900
Q2343-BX418	23:46:18.57	12:47:47.38	2.3052	23.99	-0.05	0.37	21.88	1.76	5 × 900
Q2343-BX429	23:46:25.25	12:47:51.20	2.1751	25.12	0.30	0.85	21.88	1.97	4 × 900
Q2343-BX435	23:46:26.36	12:47:55.06	2.1119	24.23	0.38	1.03	20.38	1.75	4 × 900
Q2343-BX436	23:46:09.06	12:47:56.00	2.3277	23.07	0.12	0.47	21.04	0.73	4 × 900
Q2343-BX442	23:46:19.36	12:47:59.69	2.1760	24.48	0.40	1.14	19.85	2.36	5 × 900
Q2343-BX461	23:46:32.96	12:48:08.15	2.5662	24.40	0.44	0.90	21.67	1.99	4 × 900
Q2343-BX474	23:46:32.88	12:48:14.08	2.2257	24.42	0.31	1.15	20.56	1.73	4 × 900
Q2343-BX480	23:46:21.90	12:48:15.61	2.2313	23.77	0.29	1.03	20.44	1.92	4 × 900
Q2343-BX493	23:46:14.46	12:48:21.64	2.3396	23.63	0.28	0.78	21.65	0.75	2 × 900
Q2343-BX513	23:46:11.13	12:48:32.14	2.1092[d]	23.93	0.20	0.41	20.10	1.87	4 × 900
Q2343-BX529	23:46:09.72	12:48:40.33	2.1129	24.42	0.20	0.91	21.41	1.52	2 × 900
Q2343-BX537	23:46:25.55	12:48:44.54	2.3396	24.44	0.23	0.72	21.43	1.65	4 × 900
Q2343-BX587	23:46:29.18	12:49:03.34	2.2430	23.47	0.32	1.12	20.12	1.82	3 × 900
Q2343-BX599	23:46:13.85	12:49:11.31	2.0116	23.50	0.10	0.81	20.40	1.19	4 × 900
Q2343-BX601	23:46:20.40	12:49:12.91	2.3769	23.48	0.22	0.75	20.55	1.50	4 × 900
Q2343-BX610	23:46:09.43	12:49:19.21	2.2094	23.58	0.34	0.75	19.21	2.24	2 × 900
Q2343-BX660	23:46:29.43	12:49:45.54	2.1735	24.36	-0.09	0.45	20.98	2.26	4 × 900
Q2343-MD59	23:46:26.90	12:47:39.87	2.0116	24.99	0.20	1.47	20.14	2.59	4 × 900
Q2343-MD62	23:46:27.23	12:47:43.48	2.1752	25.29	0.21	1.23	21.45	2.23	4 × 900
Q2343-MD80	23:46:10.79	12:48:33.24	2.0138	24.81	0.09	1.33	21.38	1.15	4 × 900
Q2346-BX120	23:48:26.30	00:20:33.16	2.2664	25.08	0.02	0.91	...	...	4 × 900
Q2346-BX220	23:48:46.10	00:22:20.95	1.9677	23.57	0.29	0.90	20.82	...	4 × 900
Q2346-BX244	23:48:09.61	00:22:36.18	1.6465	24.54	0.41	1.16	...	...	4 × 900
Q2346-BX404	23:48:21.40	00:24:43.07	2.0282	23.39	0.18	0.41	20.05	...	5 × 900
Q2346-BX405	23:48:21.22	00:24:45.46	2.0300	23.36	0.08	0.60	20.27	...	5 × 900
Q2346-BX416	23:48:18.21	00:24:55.30	2.2404	23.49	0.40	0.75	20.30	...	3 × 900
Q2346-BX482	23:48:12.97	00:25:46.34	2.2569	23.32	0.22	0.90	...	...	4 × 900
SSA22a-MD41[c]	22:17:39.97	00:17:11.04	2.1713	23.31	0.19	1.31	...	...	15 × 720
West-BM115	14:17:37.57	52:27:05.42	1.6065	23.41	0.28	0.36	...	...	10 × 900
West-BX600	14:17:15.55	52:36:15.64	2.1607	23.94	0.10	0.46	...	...	5 × 900

Table 4.1—Continued

Object	R.A.	Dec.	$z_{H\alpha}$	$\mathcal{R}^a$	$G - \mathcal{R}^a$	$U_n - G^a$	K_s^b	$J - K_s^b$	Exposure time (s)

[a]U_n, G, and $\mathcal{R}$ magnitudes are AB.

[b]J and K_s magnitudes are Vega.

[c]Observed with the ISAAC spectrograph on the VLT; previously discussed by Erb et al. (2003).

[d]Q2343-BX513 was observed a second time with a different position angle, and yielded an Hα redshift $z_{H\alpha} = 2.1079$.

Table 4.2.   Near-IR Imaging

Field	Band	Exposure time (hrs)	Depth[a]
GOODS-N	J	13.0	24.1
	K_s	10.3	22.6
Q1623	J	9.8	23.8
	K_s	11.2	22.3
Q1700	J	10.7	24.0
	K_s	11.0	22.2
Q2343	J	10.7	24.0
	K_s	12.1	22.3
Q2346	K_s	2.6	21.2

[a]Approximate 3σ image depth, in Vega magnitudes.

Table 4.3.   Results of SED Fitting

Object	τ (Myr)	$E(B-V)^{\mathrm{a}}$	Age[b] (Myr)	SFR[c] ($M_\odot$ yr^{-1})	$M_\star{}^{\mathrm{d}}$ (10^{10} $M_\odot$)
HDF-BMZ1156	500	0.000	1900	4	9.6
HDF-BX1277	const	0.095	321	15	0.5
HDF-BX1303	const	0.100	1139	7	0.8
HDF-BX1311	const	0.105	255	34	0.9
HDF-BX1322	const	0.085	360	19	0.7
HDF-BX1332	const	0.290	15	159	0.2
HDF-BX1368	const	0.160	404	37	1.5
HDF-BX1376	const	0.070	227	9	0.2
HDF-BX1388	const	0.265	3000	23	7.0
HDF-BX1397	const	0.150	1015	17	1.7
HDF-BX1409	const	0.290	1015	27	2.7
HDF-BX1439	const	0.175	2100	21	4.5
HDF-BX1479	const	0.110	806	13	1.0
HDF-BX1564	100	0.065	360	9	3.3
HDF-BX1567	50	0.050	227	5	2.3
HDF-BX305	const	0.285	571	53	3.0
Q1623-BX376	const	0.175	64	80	0.5
Q1623-BX428	50	0.000	255	1	1.1
Q1623-BX429	const	0.120	227	23	0.5
Q1623-BX432	const	0.060	1278	6	0.8
Q1623-BX447	500	0.050	1434	5	4.4
Q1623-BX449	const	0.110	2600	9	2.4
Q1623-BX452	const	0.195	3000	14	4.3
Q1623-BX453	const	0.275	454	107	4.9
Q1623-BX455	const	0.265	15	58	0.09
Q1623-BX458	const	0.165	571	55	3.1
Q1623-BX472	const	0.130	3000	11	3.2
Q1623-BX502	const	0.220	6	72	0.04
Q1623-BX511	const	0.235	571	13	0.8
Q1623-BX513	const	0.145	454	46	2.1
Q1623-BX516	const	0.145	1800	28	5.1
Q1623-BX522	const	0.180	2600	24	6.2
Q1623-BX528	const	0.175	2750	44	12.2
Q1623-BX543	const	0.305	8	528	0.4
Q1623-BX586	const	0.195	1434	17	2.5
Q1623-BX599	const	0.125	1900	35	6.7
Q1623-BX663	1000	0.135	2000	21	13.2
Q1623-MD107	const	0.060	1015	4	0.4
Q1623-MD66	const	0.235	905	43	3.9
Q1700-BX490	const	0.285	10	448	0.4
Q1700-BX505	const	0.270	1800	20	3.6
Q1700-BX523	const	0.260	255	42	1.1
Q1700-BX530	50	0.045	203	6	1.8
Q1700-BX536	50	0.115	180	15	2.8
Q1700-BX561	500	0.130	1609	10	11.5
Q1700-BX581	const	0.215	35	70	0.2
Q1700-BX681	const	0.315	10	628	0.6
Q1700-BX691	1000	0.125	2750	5	7.6
Q1700-BX717	const	0.090	509	8	0.4

Table 4.3—Continued

Object	τ (Myr)	$E(B-V)^{\mathrm{a}}$	Age[b] (Myr)	SFR[c] ($\mathrm{M_\odot\ yr^{-1}}$)	$M_\star$[d] ($10^{10}\ \mathrm{M_\odot}$)
Q1700-BX759	const	0.230	640	37	2.4
Q1700-BX794	const	0.130	454	25	1.1
Q1700-BX917	200	0.040	806	4	4.0
Q1700-MD103	const	0.305	1015	65	6.6
Q1700-MD109	const	0.175	2200	8	1.7
Q1700-MD154	const	0.335	128	347	4.4
Q1700-MD174	1000	0.195	2400	24	23.6
Q1700-MD69	2000	0.275	2750	31	18.6
Q1700-MD94	const	0.500	719	213	15.3
Q2343-BM133	const	0.115	143	35	0.5
Q2343-BX163	const	0.050	1434	9	1.3
Q2343-BX169	const	0.125	203	46	0.9
Q2343-BX182	const	0.100	180	23	0.4
Q2343-BX236	const	0.085	1680	13	2.1
Q2343-BX336	const	0.210	321	58	1.9
Q2343-BX341	const	0.210	102	50	0.5
Q2343-BX378	const	0.165	255	11	0.3
Q2343-BX389	const	0.250	2750	22	6.1
Q2343-BX390	const	0.150	404	17	0.7
Q2343-BX391	const	0.195	64	25	0.2
Q2343-BX418	const	0.035	81	12	0.1
Q2343-BX429	const	0.185	321	12	0.4
Q2343-BX435	const	0.225	1434	30	4.4
Q2343-BX436	const	0.070	321	33	1.1
Q2343-BX442	2000	0.225	2750	25	14.7
Q2343-BX461	const	0.250	15	86	0.1
Q2343-BX474	const	0.215	2750	26	7.2
Q2343-BX480	const	0.165	905	33	3.0
Q2343-BX493	const	0.255	6	220	0.1
Q2343-BX513	const	0.135	3000	20	5.9
Q2343-BX529	const	0.145	404	14	0.6
Q2343-BX537	const	0.130	571	15	0.8
Q2343-BX587	const	0.180	719	49	3.5
Q2343-BX599	const	0.100	1609	21	3.3
Q2343-BX601	const	0.125	640	36	2.3
Q2343-BX610	1000	0.155	2100	32	23.2
Q2343-BX660	const	0.010	2750	5	1.4
Q2343-MD59	2000	0.200	3000	11	7.6
Q2343-MD62	const	0.150	2750	7	1.9
Q2343-MD80	50	0.020	255	1	0.6
Q2346-BX220	50	0.055	227	4	1.7
Q2346-BX404	const	0.095	1800	22	4.0
Q2346-BX405	100	0.010	321	7	1.6
Q2346-BX416	const	0.195	454	55	2.5

[a]Typical uncertainty $\langle \sigma_{E(B-V)}/E(B-V) \rangle = 0.7$.

[b]Typical uncertainty $\langle \sigma_{\mathrm{Age}}/\mathrm{Age} \rangle = 0.5$.

[c]Typical uncertainty $\langle \sigma_{\mathrm{SFR}}/\mathrm{SFR} \rangle = 0.6$.

[d]Typical uncertainty $\langle \sigma_{M_\star}/M_\star \rangle = 0.4$.

Table 4.4.   Kinematics

Object	$z_{\mathrm{H}\alpha}$	z_{abs}[a]	$z_{\mathrm{Ly}\alpha}$[b]	$r_{\mathrm{H}\alpha}$[c] (kpc)	σ[d] (km s^{-1})	v_c[e] (km s^{-1})	M_{dyn}[f] (10^{10} M$_\odot$)	$M_\star$[g] (10^{10} M$_\odot$)
CDFb-BN88	2.2615	...	...	3.1	96^{+13}_{-13}	...	3.3	...
HDF-BX1055	2.4899	2.4865	2.4959	1.6	< 59	...	< 0.6	...
HDF-BX1084	2.4403	2.4392	...	2.0	102^{+4}_{-4}	...	3.2	...
HDF-BX1085	2.2407	2.2381	...	1.0	...	...	...	...
HDF-BX1086	2.4435	...	...	< 0.8	95^{+13}_{-13}	...	< 1.0	...
HDF-BX1277	2.2713	2.2686	...	2.1	63^{+6}_{-7}	...	0.8	0.5
HDF-BX1303	2.3003	2.3024	2.3051	1.4	...	...	...	0.8
HDF-BX1311	2.4843	2.4804	2.4890	2.3	88^{+6}_{-6}	...	2.0	0.9
HDF-BX1322	2.4443	2.4401	2.4491	1.9	< 47	...	< 0.7	0.7
HDF-BX1332	2.2136	2.2113	...	2.3	54^{+11}_{-13}	47 ± 14	0.8	0.2
HDF-BX1368	2.4407	2.4380	2.4455	3.1	139^{+6}_{-6}	...	4.1	1.5
HDF-BX1376	2.4294	2.4266	2.4338	1.2	96^{+15}_{-16}	...	1.3	0.2
HDF-BX1388	2.0317	2.0305	...	3.3	140^{+14}_{-14}	...	7.3	7.0
HDF-BX1397	2.1328	2.1322	...	3.6	125^{+15}_{-16}	109 ± 19	6.4	1.7
HDF-BX1409	2.2452	2.2433	...	2.9	158^{+12}_{-12}	...	8.3	2.7
HDF-BX1439	2.1865	2.1854	2.1913	3.4	120^{+5}_{-5}	...	5.4	4.5
HDF-BX1479	2.3745	2.3726	2.3823	1.8	46^{+12}_{-14}	...	0.6	1.0
HDF-BX1564	2.2225	2.2219	...	6.4	99^{+11}_{-12}	...	7.3	3.3
HDF-BX1567	2.2256	2.2257	...	1.1	< 62	...	< 1.1	2.3
HDF-BX305	2.4839	2.4825	...	1.6	140^{+15}_{-16}	...	3.5	3.0
HDF-BMZ1156[h]	2.2151	...	...	< 0.8	196^{+16}_{-17}	...	< 4.4	9.6
Q0201-B13	2.1663	...	...	3.1	62^{+7}_{-7}	...	1.4	...
Q1307-BM1163	1.4105	1.4080	...	2.8	125^{+5}_{-5}	...	5.1	...
Q1623-BX151[h]	2.4393	...	...	3.5	98^{+24}_{-27}	...	3.9	...
Q1623-BX214	2.4700	2.4674	...	1.7	55^{+9}_{-10}	...	0.6	...
Q1623-BX215	2.1814	2.1819	...	2.1	70^{+10}_{-11}	...	1.2	...
Q1623-BX252	2.3367	...	...	< 0.8	...	...	...	...
Q1623-BX274	2.4100	2.4081	2.4130	2.0	121^{+6}_{-6}	...	3.4	...
Q1623-BX344	2.4224	...	...	1.7	92^{+6}_{-7}	...	1.7	...
Q1623-BX366	2.4204	2.4169	...	3.5	103^{+26}_{-29}	...	4.4	...
Q1623-BX376	2.4085	2.4061	2.4153	2.9	261^{+36}_{-36}	...	22.9	0.5
Q1623-BX428	2.0538	2.0514	2.0594	0.6	...	...	...	1.1
Q1623-BX429	2.0160	2.0142	...	2.8	57^{+15}_{-18}	...	1.1	0.5
Q1623-BX432	2.1817	...	...	1.9	54^{+10}_{-11}	...	0.7	0.8
Q1623-BX447	2.1481	2.1478	...	2.2	174^{+10}_{-10}	160 ± 22	7.8	4.4
Q1623-BX449	2.4188	...	...	< 0.8	< 72	...	< 1.5	2.4
Q1623-BX452	2.0595	2.0595	...	4.0	129^{+13}_{-13}	...	8.1	4.3
Q1623-BX453	2.1816	2.1724	2.1838	1.7	61^{+3}_{-3}	...	0.7	4.9
Q1623-BX455	2.4074	2.4066	...	2.1	187^{+10}_{-10}	...	8.5	0.09
Q1623-BX458	2.4194	2.4174	...	3.5	160^{+18}_{-18}	116 ± 18	10.5	3.1
Q1623-BX472	2.1142	2.1144	...	2.5	110^{+7}_{-7}	110 ± 12	3.6	3.2
Q1623-BX502	2.1558	2.1549	2.1600	2.6	75^{+5}_{-5}	...	1.7	0.04
Q1623-BX511	2.2421	...	...	2.6	152^{+23}_{-24}	80 ± 18	7.1	0.8
Q1623-BX513	2.2473	2.2469	2.2525	1.4	...	...	...	2.1
Q1623-BX516	2.4236	2.4217	...	1.1	114^{+10}_{-11}	...	1.7	5.1
Q1623-BX522	2.4757	2.4742	...	2.6	< 46	...	< 0.6	6.2
Q1623-BX528	2.2682	2.2683	...	3.3	142^{+9}_{-9}	76 ± 12	7.5	12.2
Q1623-BX543	2.5211	2.5196	...	3.1	148^{+16}_{-17}	...	7.8	0.4
Q1623-BX586	2.1045	...	...	2.2	124^{+12}_{-13}	...	3.8	2.5

Table 4.4—Continued

Object	$z_{H\alpha}$	z_{abs}[a]	$z_{Ly\alpha}$[b]	$r_{H\alpha}$[c] (kpc)	σ[d] (km s^{-1})	v_c[e] (km s^{-1})	M_{dyn}[f] (10^{10} M$_\odot$)	$M_\star$[g] (10^{10} M$_\odot$)
Q1623-BX599	2.3304	2.3289	2.3402	2.4	162^{+5}_{-5}	...	7.2	6.7
Q1623-BX663[h]	2.4333	2.4296	2.4353	2.9	132^{+8}_{-9}	...	5.8	13.2
Q1623-MD107	2.5373	...	...	1.5	< 43	...	< 0.6	0.4
Q1623-MD66	2.1075	2.1057	...	2.2	120^{+3}_{-3}	...	3.7	3.9
Q1700-BX490	2.3960	2.3969	2.4043	2.3	110^{+6}_{-6}	...	3.2	0.4
Q1700-BX505	2.3089	...	...	3.1	120^{+13}_{-14}	...	5.2	3.6
Q1700-BX523	2.4756	...	...	3.2	130^{+17}_{-17}	...	6.2	1.1
Q1700-BX530	1.9429	1.9411	...	2.6	< 37	...	< 0.4	1.8
Q1700-BX536	1.9780	...	...	3.3	89^{+9}_{-10}	...	3.0	2.8
Q1700-BX561	2.4332	2.4277	...	< 0.8	...	...	...	11.5
Q1700-BX581	2.4022	2.3984	...	1.8	...	...	...	0.2
Q1700-BX681	1.7396	1.7398	1.7467	3.6	...	...	...	0.6
Q1700-BX691	2.1895	...	...	2.8	170^{+9}_{-9}	220 ± 14	9.4	7.6
Q1700-BX717	2.4353	...	2.4376	2.0	< 47	...	< 0.7	0.4
Q1700-BX759	2.4213	...	...	1.0	...	...	...	2.4
Q1700-BX794	2.2473	...	...	2.0	80^{+9}_{-9}	...	1.5	1.1
Q1700-BX917	2.3069	2.3027	...	4.4	99^{+8}_{-9}	...	5.1	4.0
Q1700-MD69	2.2883	2.288	...	3.9	155^{+10}_{-11}	...	10.8	18.6
Q1700-MD94[h]	2.3362	...	...	4.0	730^{+55}_{-55}	...	238.6	15.3
Q1700-MD103	2.3148	...	...	3.5	75^{+14}_{-16}	100 ± 19	2.3	6.6
Q1700-MD109	2.2942	...	...	1.7	93^{+16}_{-17}	...	1.7	1.7
Q1700-MD154	2.6291	...	...	3.4	57^{+23}_{-33}	...	1.3	4.4
Q1700-MD174[h]	2.3423	...	...	1.5	444^{+33}_{-33}	...	35.3	23.6
Q2343-BM133	1.4774	1.4769	...	3.2	55^{+4}_{-4}	...	1.1	0.5
Q2343-BM181	1.4951	1.4952	...	3.2	39^{+17}_{-30}	45 ± 18	0.6	...
Q2343-BX163	2.1213	...	...	1.7	...	...	...	1.3
Q2343-BX169	2.2094	2.2105	2.2173	1.3	...	...	...	0.9
Q2343-BX182	2.2879	2.2857	2.2909	1.1	...	...	...	0.4
Q2343-BX236	2.4348	2.4304	2.4372	2.0	148^{+27}_{-28}	...	5.1	2.1
Q2343-BX336	2.5439	2.5448	2.5516	2.1	140^{+22}_{-23}	...	4.8	1.9
Q2343-BX341	2.5749	2.5715	...	1.0	...	...	...	0.5
Q2343-BX378	2.0441	...	...	3.1	...	...	...	0.3
Q2343-BX389	2.1716	2.1722	...	4.2	111^{+5}_{-5}	71 ± 10	6.0	6.1
Q2343-BX390	2.2313	2.2290	...	4.5	78^{+14}_{-16}	...	3.2	0.7
Q2343-BX391	2.1740	2.1714	...	2.5	...	...	...	0.2
Q2343-BX418	2.3052	2.3030	2.3084	1.5	66^{+4}_{-4}	...	0.7	0.1
Q2343-BX429	2.1751	...	...	3.9	51^{+11}_{-13}	...	1.2	0.4
Q2343-BX435	2.1119	2.1088	2.1153	3.7	60^{+8}_{-9}	...	1.5	4.4
Q2343-BX436	2.3277	2.3253	2.3315	3.8	63^{+7}_{-8}	...	1.8	1.1
Q2343-BX442	2.1760	...	...	4.4	132^{+6}_{-6}	...	8.5	14.7
Q2343-BX461	2.5662	2.5649	2.5759	3.2	139^{+15}_{-16}	...	7.2	0.1
Q2343-BX474	2.2257	2.2263	...	2.9	84^{+8}_{-8}	...	2.4	7.2
Q2343-BX480	2.2313	2.2297	2.2352	1.4	< 37	...	< 0.4	3.0
Q2343-BX493	2.3396	2.3375	2.3447	2.7	155^{+26}_{-28}	...	7.6	0.1
Q2343-BX513[i]	2.1092	2.1090	2.114	1.7	150^{+6}_{-6}	...	4.4	5.9
Q2343-BX529	2.1129	2.1116	2.1190	< 0.8	...	...	...	0.6
Q2343-BX537	2.3396	...	...	2.2	...	...	...	0.8
Q2343-BX587	2.2430	2.2382	...	3.2	94^{+9}_{-10}	...	3.2	3.5
Q2343-BX599	2.0116	2.0112	...	3.0	77^{+14}_{-15}	...	2.1	3.3

Table 4.4—Continued

Object	$z_{H\alpha}$	z_{abs}[a]	$z_{Ly\alpha}$[b]	$r_{H\alpha}$[c] (kpc)	σ[d] (km s^{-1})	v_c[e] (km s^{-1})	M_{dyn}[f] (10^{10} M$_\odot$)	$M_\star$[g] (10^{10} M$_\odot$)
Q2343-BX601	2.3769	2.3745	2.3823	2.2	105^{+8}_{-8}	...	2.8	2.3
Q2343-BX610	2.2094	2.2083	2.2129	3.5	96^{+6}_{-7}	...	3.8	23.2
Q2343-BX660	2.1735	2.1709	2.1771	2.6	< 40	...	< 0.5	1.4
Q2343-MD59	2.0116	2.0107	...	2.3	...	...	...	7.6
Q2343-MD62	2.1752	2.1740	...	2.8	79^{+18}_{-20}	...	2.0	1.9
Q2343-MD80	2.0138	2.0116	...	1.1	74^{+11}_{-11}	...	0.7	0.6
Q2346-BX120	2.2664	...	...	2.6	62^{+8}_{-8}	40 ± 10	1.2	...
Q2346-BX220	1.9677	1.9664	...	2.7	143^{+9}_{-9}	...	6.5	1.7
Q2346-BX244	1.6465	1.6462	1.6516	3.2	42^{+18}_{-28}	...	0.7	...

Table 4.4—Continued

Object	$z_{H\alpha}$	z_{abs}[a]	$z_{Ly\alpha}$[b]	$r_{H\alpha}$[c] (kpc)	σ[d] (km s^{-1})	v_c[e] (km s^{-1})	M_{dyn}[f] (10^{10} M$_\odot$)	$M_\star$[g] (10^{10} M$_\odot$)
Q2346-BX404	2.0282	2.0270	2.0348	1.5	102^{+2}_{-2}	...	1.9	4.0
Q2346-BX405	2.0300	2.0298	2.0358	2.4	50^{+3}_{-3}	...	0.7	1.6
Q2346-BX416	2.2404	2.2407	...	1.8	126^{+8}_{-9}	...	3.3	2.5
Q2346-BX482	2.2569	2.2575	...	4.1	133^{+3}_{-3}	...	8.6	...
SSA22a-MD41	2.1713	...	...	4.0	107^{+4}_{-4}	150 ± 16	5.1	...
West-BM115	1.6065	1.6060	...	2.3	128^{+10}_{-10}	...	4.3	...
West-BX600	2.1607	...	...	3.4	181^{+11}_{-11}	210 ± 13	13.4	...

[a]Vacuum redshift of the UV insterstellar absorption lines. We give a value only when the S/N of the spectrum is sufficient for a precise measurement.

[b]Vacuum redshift of the Lyα emission line, when present.

[c]Approximate spatial extent of the Hα emission, FWHM/2.4, after subtraction of the seeing in quadrature.

[d]Velocity dispersion of the Hα emission line.

[e]For tilted emission lines, the velocity shear $(v_{max} - v_{min})/2$, where v_{max} and v_{min} are with respect to the systemic redshift.

[f]Dynamical mass $M_{dyn} = 5\sigma^2 r_{H\alpha}/G$.

[g]Stellar mass, from SED modeling.

[h]AGN

[i]Q2343-BX513 was observed twice with NIRSPEC, with position angles differing by 9°. The first observation yielded $z_{H\alpha} = 2.1079$ and $\sigma = 58$ km s^{-1}, and the second $z_{H\alpha} = 2.1092$ and $\sigma = 150$ km s^{-1}. It also has two Lyα emission redshifts, $z_{lya} = 2.106$ and $z_{lya} = 2.114$.

Table 4.5.　Star Formation Rates

Object	$E(B-V)$[a]	$F_{H\alpha}$[b]	$L_{H\alpha}$[c]	Uncorrected $SFR_{H\alpha}$[d]	Corrected $SFR_{H\alpha}$[e]	m_{1500}[f]	$Log(L_{1500})$[g]	Uncorrected SFR_{UV}[h]	Corrected SFR_{UV}[i]	SFR_{fit}[j]	$W_{H\alpha}$[k]
CDFb-BN88	0.149	2.6	1.0	4	14	23.43	29.27	14	60	...	...
HDF-BX1055	0.103	2.6	1.3	6	15	24.33	28.98	8	21	...	...
HDF-BX1084	0.120	7.3	3.4	15	44	23.50	29.30	16	51	...	...
HDF-BX1085	0.171	1.1	0.4	2	6	24.83	28.70	4	20	...	...
HDF-BX1086	0.196	1.8	0.8	4	14	25.05	28.68	4	26	...	...
HDF-BMZ1156	0.000	5.3	2.1	9	18	24.01	29.04	8	8	4	116
HDF-BX1277	0.095	2.6	1.0	5	12	24.83	28.72	4	10	15	308
HDF-BX1303	0.100	8.0	3.9	17	47	23.50	29.31	16	43	7	101
HDF-BX1311	0.105	2.0	0.9	4	11	24.03	29.09	10	27	34	197
HDF-BX1322	0.085	4.4	1.6	7	19	23.96	29.04	8	19	19	68
HDF-BX1332	0.290	8.8	4.1	18	88	24.09	29.06	9	159	159	132
HDF-BX1368	0.160	2.2	1.0	4	15	24.49	28.90	6	30	37	266
HDF-BX1376	0.070	5.8	1.8	8	19	24.82	28.63	3	6	9	265
HDF-BX1388	0.265	5.3	1.8	8	36	24.26	28.89	6	73	23	90
HDF-BX1397	0.150	8.5	3.2	14	45	25.15	28.57	3	12	17	207
HDF-BX1409	0.290	8.8	3.2	14	67	24.16	28.95	7	108	27	145
HDF-BX1439	0.175	2.5	1.1	5	16	24.55	28.86	6	31	21	107
HDF-BX1479	0.110	8.6	3.2	14	40	23.55	29.21	13	36	13	126
HDF-BX1564	0.065	4.0	1.5	7	16	23.68	29.15	11	21	9	93
HDF-BX1567	0.050	4.2	2.1	9	21	25.07	28.68	4	6	5	72
HDF-BX305	0.285	5.4	2.0	9	42	24.61	28.78	5	71	53	67
Q0201-B13	0.003	2.4	0.8	4	8	23.36	29.26	14	15	...	...
Q1307-BM1163	0.178	28.7	3.5	15	53	22.21	29.38	19	99	...	...
Q1623-BX151	0.059	3.5	1.6	7	17	24.74	28.80	5	9	...	...
Q1623-BX214	0.182	5.3	2.6	11	39	24.45	28.92	7	40	...	...
Q1623-BX215	0.134	4.8	1.7	8	22	24.71	28.73	4	15	...	...
Q1623-BX252	0.031	1.2	0.5	2	5	25.13	28.61	3	4	...	...
Q1623-BX274	0.119	9.5	4.3	19	54	23.48	29.29	15	50	...	...
Q1623-BX344	0.189	17.1	7.9	35	123	24.81	28.77	5	30	...	...
Q1623-BX366	0.200	7.9	3.6	16	58	24.25	28.99	8	55	...	...
Q1623-BX376	0.175	5.3	2.4	11	36	23.55	29.27	14	81	80	183
Q1623-BX428	0.000	2.7	0.8	4	7	24.08	28.93	7	7	1	84
Q1623-BX429	0.120	5.1	1.5	7	19	23.75	29.05	9	26	23	219
Q1623-BX432	0.060	5.4	1.9	8	20	24.68	28.74	4	8	6	427
Q1623-BX447	0.050	5.6	1.9	8	20	24.65	28.74	4	7	5	154
Q1623-BX449	0.110	3.5	1.6	7	20	25.06	28.67	4	11	9	196
Q1623-BX452	0.195	4.4	1.4	6	22	24.93	28.60	3	19	14	121

Table 4.5—Continued

Object	$E(B-V)$[a]	$F_{H\alpha}$[b]	$L_{H\alpha}$[c]	Uncorrected $SFR_{H\alpha}$[d]	Corrected $SFR_{H\alpha}$[e]	m_{1500}[f]	$Log(L_{1500})$[g]	Uncorrected SFR_{UV}[h]	Corrected SFR_{UV}[i]	SFR_{fit}[j]	$W_{H\alpha}$[k]
Q1623-BX453	0.275	13.8	4.9	22	100	23.86	29.07	9	123	107	187
Q1623-BX455	0.265	18.8	8.6	38	169	25.15	28.63	3	45	58	1172[l]
Q1623-BX458	0.165	4.3	2.0	9	29	23.69	29.21	13	65	55	102
Q1623-BX472	0.130	3.9	1.3	6	17	24.74	28.69	4	13	11	135
Q1623-BX502	0.220	13.2	4.6	20	79	24.57	28.77	5	37	72	1536[l]
Q1623-BX511	0.235	3.4	1.3	6	23	25.79	28.32	2	15	13	325
Q1623-BX513	0.145	3.3	1.3	6	17	23.51	29.23	13	53	46	59
Q1623-BX516	0.145	5.2	2.4	10	33	24.24	28.99	8	32	28	112
Q1623-BX522	0.180	2.8	1.4	6	21	24.81	28.78	5	28	24	79
Q1623-BX528	0.175	7.7	3.0	13	46	23.81	29.12	10	55	44	94
Q1623-BX543	0.305	8.6	4.4	19	98	23.55	29.30	16	336	528	229
Q1623-BX586	0.195	5.1	1.7	7	27	24.90	28.62	3	20	17	192
Q1623-BX599	0.125	18.1	7.6	33	98	23.66	29.20	12	42	35	303
Q1623-BX663	0.135	8.2	3.8	17	50	24.38	28.94	7	26	21	112
Q1623-MD107	0.060	3.7	1.9	8	20	25.47	28.54	3	5	4	858
Q1623-MD66	0.235	19.7	6.5	28	116	24.32	28.86	6	50	43	482
Q1700-BX490	0.285	17.7	8.0	35	166	23.24	29.39	19	313	448	310
Q1700-BX505	0.270	3.6	1.5	6	29	25.62	28.41	2	27	20	121
Q1700-BX523	0.260	4.7	2.3	10	44	24.97	28.72	4	55	42	171
Q1700-BX530	0.045	12.2	3.3	14	33	23.26	29.22	13	19	6	208
Q1700-BX536	0.115	11.3	3.2	14	40	23.21	29.25	14	40	15	150
Q1700-BX561	0.130	1.9	0.9	4	11	24.84	28.76	4	16	10	22
Q1700-BX581	0.215	4.0	1.8	8	30	24.15	29.02	8	69	70	124
Q1700-BX681	0.315	6.3	1.3	6	30	22.23	29.54	27	427	628	52
Q1700-BX691	0.125	7.7	2.8	12	36	25.55	28.39	2	6	5	257
Q1700-BX717	0.090	3.8	1.8	8	20	24.98	28.70	4	10	8	410
Q1700-BX759	0.230	1.3	0.6	3	11	24.79	28.77	5	45	37	57
Q1700-BX794	0.130	6.8	2.6	12	34	23.95	29.05	9	31	25	183
Q1700-BX917	0.040	7.4	3.0	13	30	24.71	28.77	5	7	4	117
Q1700-MD103	0.305	8.2	3.4	15	76	24.69	28.78	5	90	65	120
Q1700-MD109	0.175	2.8	1.1	5	17	25.72	28.36	2	10	8	246
Q1700-MD154	0.335	4.1	2.3	10	56	23.96	29.17	12	359	347	40
Q1700-MD174	0.195	8.9	3.8	17	60	24.88	28.71	4	27	24	125
Q1700-MD69	0.275	7.5	3.0	13	61	25.22	28.56	3	40	31	122
Q1700-MD94	0.500	12.9	5.4	24	219	25.66	28.40	2	253	213	146
Q2343-BM133	0.115	28.7	3.9	17	49	22.78	29.19	12	36	35	2245
Q2343-BM181	0.134	3.4	0.5	2	6	25.18	28.24	1	5	...	...

Table 4.5—Continued

Object	$E(B-V)$[a]	$F_{H\alpha}$[b]	$L_{H\alpha}$[c]	Uncorrected $SFR_{H\alpha}$[d]	Corrected $SFR_{H\alpha}$[e]	m_{1500}[f]	$Log(L_{1500})$[g]	Uncorrected SFR_{UV}[h]	Corrected SFR_{UV}[i]	SFR_{fit}[j]	$W_{H\alpha}$[k]
Q2343-BX163	0.050	2.2	0.7	3	7	24.06	28.97	7	12	9	127
Q2343-BX169	0.125	4.7	1.7	8	22	23.30	29.30	16	51	46	152
Q2343-BX182	0.100	2.4	1.0	4	11	23.88	29.10	10	26	23	168
Q2343-BX236	0.085	3.1	1.4	6	16	24.42	28.93	7	15	13	150
Q2343-BX336	0.210	4.3	2.2	10	38	24.31	29.00	8	66	58	133
Q2343-BX341	0.210	4.0	2.1	9	36	24.59	28.90	6	52	50	231
Q2343-BX378	0.165	4.5	1.4	6	20	25.06	28.54	3	12	11	606
Q2343-BX389	0.250	12.0	4.2	19	80	25.13	28.56	3	30	22	253
Q2343-BX390	0.150	4.9	1.9	8	26	24.60	28.79	5	20	17	293
Q2343-BX391	0.195	4.2	1.5	6	24	24.51	28.80	5	31	25	537
Q2343-BX418	0.035	8.0	3.3	14	32	23.94	29.08	9	13	12	1639
Q2343-BX429	0.185	4.8	1.7	8	27	25.42	28.44	2	12	12	632
Q2343-BX435	0.225	8.1	2.7	12	47	24.61	28.74	4	35	30	200
Q2343-BX436	0.070	7.2	3.0	13	33	23.19	29.38	19	37	33	345
Q2343-BX442	0.225	7.2	2.5	11	44	24.48	28.82	5	43	25	98
Q2343-BX461	0.250	7.0	3.7	16	70	24.84	28.80	5	62	86	760
Q2343-BX474	0.215	5.0	1.9	8	32	24.73	28.73	4	33	26	133
Q2343-BX480	0.165	3.0	1.1	5	16	24.06	29.00	8	38	33	67
Q2343-BX493	0.255	5.3	2.2	10	43	23.91	29.10	10	118	220	497
Q2343-BX513	0.135	10.1	3.3	15	44	24.13	28.93	7	24	20	192
Q2343-BX529	0.145	3.5	1.2	5	16	24.62	28.74	4	17	14	230
Q2343-BX537	0.130	5.2	2.2	10	29	24.67	28.80	5	17	15	365
Q2343-BX587	0.180	5.5	2.1	9	32	23.79	29.12	10	57	49	95
Q2343-BX599	0.100	4.5	1.3	6	16	23.60	29.11	10	25	21	107
Q2343-BX601	0.125	7.4	3.3	14	42	23.70	29.20	12	42	36	199
Q2343-BX610	0.155	8.1	3.0	13	42	23.92	29.05	9	38	32	59
Q2343-BX660	0.010	9.4	3.3	15	30	24.27	28.90	6	7	5	488
Q2343-MD59	0.200	2.9	0.8	4	14	24.99	28.55	3	18	11	52
Q2343-MD62	0.150	2.3	0.8	4	11	25.50	28.41	2	8	7	143
Q2343-MD80	0.020	3.2	0.9	4	9	24.81	28.63	3	4	1	206
Q2346-BX120	0.005	5.3	2.1	9	19	25.10	28.60	3	3	...	...
Q2346-BX220	0.055	10.3	2.9	13	30	23.86	28.99	8	13	4	482
Q2346-BX244	0.300	5.4	1.0	4	21	23.49	29.00	8	149	...	...
Q2346-BX404	0.095	13.9	4.2	18	49	23.57	29.13	10	25	22	273
Q2346-BX405	0.010	14.0	4.2	18	38	23.44	29.18	12	13	7	358
Q2346-BX416	0.195	12.1	4.6	20	73	23.89	29.08	9	60	55	287
Q2346-BX482	0.112	11.2	4.4	19	54	23.54	29.22	13	38	...	...

Table 4.5—Continued

Object	$E(B-V)^a$	$F_{H\alpha}^b$	$L_{H\alpha}^c$	Uncorrected $SFR_{H\alpha}^d$	Corrected $SFR_{H\alpha}^e$	m_{1500}^f	$\mathrm{Log}(L_{1500})^g$	Uncorrected SFR_{UV}^h	Corrected SFR_{UV}^i	SFR_{fit}^j	$W_{H\alpha}^k$
SSA22a-MD41	0.096	7.9	2.8	12	33	23.50	29.21	13	31	...	...
West-BM115	0.225	5.9	1.0	4	17	24.05	28.75	4	40	...	...
West-BX600	0.047	6.3	2.2	10	22	24.04	28.99	8	12	...	...

[a]$E(B-V)$ inferred from SED fitting when K-band photometry is present (indicated by a value in column 11, the SFR from SED fitting), and calculated from the $G-\mathcal{R}$ color assuming an SED with constant star formation and an age of 700 Myr otherwise.

[b]Flux of Hα line, in units of 10^{-17} erg s^{-1} cm^{-2}.

[c]Hα luminosity, in units of 10^{42} erg s^{-1}.

[d]SFR derived from observed Hα flux, in M$_\odot$ yr^{-1}.

[e]SFR derived from Hα flux after correcting for extinction and slit losses as described in the text, in M$_\odot$ yr^{-1}.

[f]Observed magnitude at ~ 1500 Å; G-band for most objects, U_n for those with $z \sim 1.5$.

[g]Rest-frame UV luminosity, log (erg s^{-1} cm^{-2} Hz^{-1}).

[h]SFR derived from uncorrected UV magnitude, in M$_\odot$ yr^{-1}.

[i]SFR derived from extinction-corrected UV magnitude, in M$_\odot$ yr^{-1}.

[j]SFR derived from SED fitting, in M$_\odot$ yr^{-1}.

[k]Hα equivalent width in Å, incorporating a factor of 2 aperture correction except where noted.

[l]Aperture correction not applied for equivalent width calculation.

Chapter 5

The Mass-Metallicity Relation at $z \gtrsim 2$*

Dawn K. Erb,[a] Alice E. Shapley,[b] Max Pettini,[c] Charles C. Steidel,[a] Naveen A. Reddy,[a] Kurt L. Adelberger[d]

[a]California Institute of Technology, MS 105–24, Pasadena, CA 91125

[b]Department of Astronomy, 601 Campbell Hall, University of California at Berkeley, Berkeley, CA 94720

[c]Institute of Astronomy, Madingley Road, Cambridge CB3 0HA, UK

[d]Carnegie Observatories, 813 Santa Barbara Street, Pasadena, CA 91101

Abstract

We use a sample of 87 rest-frame ultraviolet-selected star-forming galaxies with mean spectroscopic redshift $\langle z \rangle = 2.26 \pm 0.17$ to study the correlation between metallicity and stellar mass at high redshift. Using stellar masses determined from spectral energy distribution fitting to $U_n GRJK_s$ (and Spitzer IRAC, for 37% of the sample) photometry, we divide the sample into six bins in stellar mass, and construct six composite Hα/[N II] spectra from all of the objects in each bin. We estimate the mean oxygen abundance in each bin from the [N II]/Hα ratio, and find a monotonic increase in metallicity with increasing stellar mass, from $12 + \log(\text{O/H}) < 8.2$ for galaxies with $\langle M_\star \rangle = 2.7 \times 10^9$ M$_\odot$ to $12 + \log(\text{O/H}) = 8.6$ for galaxies with $\langle M_\star \rangle = 1.0 \times 10^{11}$ M$_\odot$. We construct a corresponding metallicity-luminosity relation by binning the galaxies according to rest-frame B magnitude, but find no significant correlation. This lack of correlation is explained by the known large variation in the rest-frame optical mass-to-light ratio at $z \sim 2$, and indicates that the correlation with stellar mass is more fundamental. On average, at a given metallicity the $z \sim 2$ galaxies are ~ 3 magnitudes brighter than local galaxies. We use the empirical relation between star formation rate density and gas density to estimate the gas fractions of the galaxies, finding an increase in gas fraction with decreasing stellar mass. Our estimates of gas fraction combined with the observed metallicities allow us to estimate the effective yield as a function of stellar mass; we find it to be constant within the uncertainties in all bins, suggesting that the simple, closed-box model of chemical evolution adequately explains our observed mass-metallicity relation. We further suggest that the range of baryonic masses spanned by our sample is too small for the *differential* effects of metal loss from winds to be significant, and that such winds may require a time comparable to the ages of the galaxies to have an appreciable effect. We conclude that the mass-metallicity relation at high redshift is driven by the increase in metallicity as the gas fraction decreases through star formation, and is later reinforced by the loss of metals through winds in low-mass galaxies.

*Based on data obtained at the W.M. Keck Observatory, which is operated as a scientific partnership among the California Institute of Technology, the University of California, and NASA, and was made possible by the generous financial support of the W.M. Keck Foundation.

5.1 Introduction

Correlations between mass and metallicity or luminosity and metallicity are well-established in nearby galaxies, ranging over orders of magnitude in mass and luminosity and spanning ~ 2 dex in chemical abundance. Lequeux et al. (1979) first observed a correlation between heavy element abundance and the total mass of galaxies; since then, most investigations have focused on the metallicity-luminosity relationship (e.g., Skillman et al. 1989; Zaritsky et al. 1994; Garnett et al. 1997; Lamareille et al. 2004; Salzer et al. 2005, to name only a few), though others have studied correlations between metallicity and rotational velocity (Zaritsky et al. 1994; Garnett 2002). The relationship between metallicity and stellar mass has recently been quantified by Tremonti et al. (2004, T04 hereafter), using a sample of $\sim 53,000$ galaxies from the Sloan Digital Sky Survey (SDSS). Such correlations can provide great insight into the process of galaxy evolution, as they allow us to study the history of star formation and gas enrichment or depletion through current, observable properties. Chemical enrichment is a record of star formation history, modulated by inflows and outflows of gas, while the stellar mass provides a more straightforward measure of the accumulated conversion of gas into stars and thus of the metals returned to the gas as the byproducts of star formation.

It has long been recognized that a correlation between stellar mass and gas-phase metallicity is a natural consequence of the conversion of gas into stars in a closed system (van den Bergh 1962; Schmidt 1963; Searle & Sargent 1972). The principal ingredients of this "simple" or "closed-box" model are the metallicity, the yield from star formation (defined as the mass of metals produced and ejected by star formation, in units of the mass that remains locked in long-lived stars and remnants), and the gas fraction. If there are no inflows or outflows of gas, the metallicity is a simple function of the yield and the gas fraction, and rises as the gas is converted into stars and enriched by star formation according to the yield. The model is subject to the further assumptions that the system is well-mixed at all times, that it begins as pure gas with primordial abundances, that stellar evolution and nucleosynthesis take place instantaneously compared to the timescale of galactic evolution (the instantaneous recycling approximation), and that the IMF and the yield in primary elements of stars of a given mass are constant.

One of the first, best-known failures of the simple model is the so-called "G-dwarf problem": the closed-box model overpredicts the number of low-metallicity stars observed in the solar neighborhood. This is one aspect of the more fundamental problem that galaxies are clearly not closed boxes. On the one hand, infall and mergers are essential aspects of galaxy formation (e.g., Pagel & Patchett 1975; Naab & Ostriker 2005). On the other, galactic-scale winds, driven by star formation, are a ubiquitous feature of starburst galaxies at low (e.g., Heckman et al. 1990; Lehnert & Heckman 1996; Martin 1999; Strickland et al. 2004, among many others) and high (Pettini et al. 2001; Shapley et al. 2003; Smail et al. 2003) redshifts. Metals are detected in the intergalactic medium (IGM; Ellison et al. 2000; Simcoe et al. 2004), and their presence is correlated with the distribution of nearby galaxies (Adelberger et al. 2003; Adelberger et al. 2005a). The potential of such supernova-powered winds to expel gas from galaxies and thus modify their chemical evolution has been known for some time; Larson (1974) showed how winds could account for the mass-metallicity relation in elliptical galaxies by preferentially ejecting metals from those with lower masses. This provides an alternative or additional explanation for the mass-metallicity correlation, and its appeal has increased in recent years as the physical evidence of feedback has multiplied and models of galaxy formation have recognized its importance (e.g., Hernquist & Springel 2003; Benson et al. 2003; Dekel & Woo 2003; Nagamine et al. 2004; Murray et al. 2005).

Motivated by these competing or complementary theories of the origin of the mass-metallicity relation, T04 use the statistical power of $\sim 53,000$ star-forming SDSS galaxies to revisit the correlation, confirming its existence over ~ 3 orders of magnitude in stellar mass and 1 dex in metallicity. Using estimates of the effective yield as a function of baryonic mass, they find strong evidence for metal depletion in low-mass galaxies, a probable signature of winds from an early starburst phase. Given the strength of the mass-metallicity correlation, and its two plausible causes, it is not unreasonable to suppose that it may be present in galaxies at high redshift, but this has been difficult to test. The strong optical emission lines usually used to determine metallicities shift into the infrared past $z \sim 1$, making the required large samples of spectra much more difficult to acquire. As luminosity can be determined with considerably greater ease than stellar mass, the first efforts focused on the luminosity-metallicity relation at high redshift (Kobulnicky & Koo 2000; Pettini et al. 2001; Shapley et al. 2004), finding that galaxies at $z > 2$ are overluminous for their metallicities when compared to local galaxies.

In this paper we study the relationships between stellar mass, luminosity, and metallicity at $z \gtrsim 2$, using a sample of 87 star-forming galaxies with [N II]/Hα spectra. We describe our sample selection, observations, and data reduction procecures in §5.2, and discuss our methods of determining stellar mass and metallicity in §5.3. In §5.4 we give our results, and in §5.5 we discuss their implications for the origin of the mass-metallicity relation. Our conclusions are presented in §5.6. We use a cosmology with $H_0 = 70$ km s^{-1} Mpc^{-1}, $\Omega_m = 0.3$, and $\Omega_\Lambda = 0.7$; in such a cosmology, the universe at $z = 2.26$ (the mean redshift of our sample) is 2.9 Gyr old, or 21% of its present age. For comparisons with solar metallicity, we use the most recent values of the solar oxygen abundance, $12 + \log(\text{O/H})_\odot = 8.66$, and the solar metal mass fraction $Z_\odot = 0.0126$ (Asplund et al. 2004).

5.2   Sample Selection, Observations, and Data Reduction

The galaxies we use in this paper are drawn from a sample of 114 galaxies with Hα spectra described in detail by Erb et al. (2005b). The galaxies are selected by their rest-frame UV colors, and their redshifts are confirmed with rest-frame UV spectra from the LRIS-B spectrograph on the 10 m Keck I telescope on Mauna Kea; an overview of the $z \sim 2$ sample is given by Steidel et al. (2004). We include all objects with Hα spectra and K-band magnitudes (most have J magnitudes as well, and 32 have also been observed at 3.6, 4.5, 5.4, and 8.0 μm with the IRAC camera on the Spitzer Space Telescope), except those with AGN signatures in either their rest-frame UV or optical spectra.

The Hα spectra were obtained between May 2002 and September 2004 with the near-IR spectrograph NIRSPEC (McLean et al. 1998) on the Keck II telescope. For the redshifts of the galaxies presented here, Hα falls in the K-band; we usually observed using the N6 filter, which spans the wavelength range 1.558–2.315 μm, and in low-dispersion mode, which provides a resolution of $R \sim 1400$. The data were reduced using standard procedures described by Erb et al. (2003, 2005b), and flux-calibrated with reference to near-IR standard stars.

The near-IR imaging was carried out with the Wide-field IR Camera (WIRC, Wilson et al. 2003) on the Palomar 5 m Hale telescope. We obtained $\sim 9 \times 9'$ images to $K_s \sim 22.5$ and $J \sim 24$ in four fields, with a typical integration time of ~ 11 hours in each band per field. Data reduction and photometry were performed as described by Erb et al. (2005b) and Shapley et al. (2005b). For a description of the mid-IR IRAC data, reductions, and photometry, see Barmby et al. (2004), Shapley et al. (2005b), and Reddy et al. (2005).

5.3   Measurements

5.3.1   Stellar Masses

Stellar masses are determined by fitting model SEDs to the $U_n G \mathcal{R} J K$ (and IRAC, when present) photometry. We use the procedure described in detail by Shapley et al. (2005b) and Erb et al. (2005b), which uses the Bruzual & Charlot (2003) population synthesis models and the Calzetti et al. (2000) extinction law. We compare the model SEDs of galaxies with a variety of ages and amounts of extinction to our observed photometry, and obtain the star formation rate and stellar mass from the normalization of the best-fit model to the data. We try models with a constant star formation (CSF) rate and models in which the star formation rate smoothly declines with time, parameterized by SFR $\propto e^{-t/\tau}$, with τ =10, 20, 100, 200, 500, 1000, 2000, and 5000 Myr. In practice, however, most star formation histories provide adequate fits to most objects, and we therefore use the CSF models unless one of the τ models provides a significantly better fit. The best-fit models for the galaxies discussed here are given by Erb et al. (2005b). We use a Chabrier (2003) IMF for the stellar masses and star formation rates.

Uncertainties in the fitting are determined from a large number of Monte Carlo simulations in which the input photometry is varied according to the photometric errors; as has often been noted for SED modeling of this sort (e.g. Papovich et al. 2001; Shapley et al. 2001, 2005b), the stellar mass is the most well-determined parameter. For the current sample, we find a mean fractional uncertainty $\langle \sigma_{M_\star}/M_\star \rangle = 0.4$.

As discussed by Papovich et al. (2001) and Shapley et al. (2005b), one limitation of such modeling is the insensitivity of the data to faint, old stellar populations, which could be obscured by current star formation and thus lead to an underestimate of the stellar mass. We have attempted to determine the magnitude of this effect by fitting a variety of two-component models to the observed SEDs, in which the light from the observed-frame K-band and redward is constrained to come from a maximally old burst while a young population is fitted to the rest-frame UV residuals. Such models make little difference to the stellar masses of already massive galaxies, since their stellar populations already approach the maximum age allowed by the age of the universe at their redshift. The masses of low-mass galaxies can be increased by an order of magnitude by such models, but the models are generally a poor fit to the SED and result in star formation rates far higher than those determined by all other indicators, with an average SFR of ~ 900 $M_\odot$ yr^{-1}. While we cannot rule out such models in individual cases, they are very unlikely to be correct on average, and we therefore consider it unlikely that we have underestimated the stellar masses of the low-mass galaxies by such a large factor. More general two-component models, in which the relative contributions from a maximally old population and a young burst are allowed to vary, increase the stellar mass by a factor of a few at most. The two-component models are described in more detail by Erb et al. (2005b).

As described below, for the purposes of determining metallicities we divide the sample into six bins by stellar mass, with 14 or 15 galaxies in each bin. The mean stellar mass in each bin ranges from 2.7×10^9 $M_\odot$ to 1.1×10^{11} $M_\odot$. The means and standard deviations of the fitted parameters for each bin are given in Table 5.1, along with the star formation rates determined from Hα luminosities. The Hα luminosities have been corrected for extinction using the Calzetti et al. (2000) extinction law and the best-fit value of $E(B-V)$ from the SED modeling, and we have applied a factor of 2 aperture correction determined from the comparison of the NIRSPEC spectra and narrowband images (see Erb et al. 2005b for details).

5.3.2 Metallicities

The most direct way to determine the abundances of metals from the observed emission line fluxes in H II regions is through the measurement of the electron temperature T_e. As the metallicity of the gas increases, the cooling through metal emission lines also increases, resulting in a decrease in T_e. The ratio of the auroral (the transition from the second lowest to the lowest excited level) and nebular (the transition from the lowest excited level to the ground state) emission lines of the same ion is highly sensitive to the electron temperature, and therefore the measurement of such pairs of lines has been the preferred method of determining abundances in H II regions. However, the auroral lines (in particular the most widely used line, [O III]λ4363) become extremely weak at metallicities above ~ 0.5 solar, and undetectable at all metallicities in the low S/N spectra of distant galaxies. In most cases, therefore, we must use the empirical "strong line" abundance indicators, which are based on the ratios of collisionally excited forbidden lines to hydrogen recombination lines. These are calibrated with reference to the T_e method or, more commonly, with detailed photoionization models. The strong line methods carry significant hazards, however. Substantial biases and offsets are observed between abundances determined with different methods, and between different calibrations of the same method (e.g., Kennicutt et al. 2003; Kobulnicky & Kewley 2004), and many of the strong line indicators are sensitive to the ionization parameter as well as metallicity. The advent of large telescopes, sensitive detectors, and spectrographs with high throughput has enabled the measurement of abundances with the T_e method in an increasingly large sample of extragalactic H II regions. These new data suggest that the most widely used indicator, $R_{23} \equiv ([O\ II]\lambda3727+[O\ III]\lambda\lambda4959,5007)/H\beta$, may systematically overestimate metallicities in the high-abundance regime ($Z \gtrsim Z_\odot$) by as much as 0.2–0.5 dex (Kennicutt et al. 2003; Garnett et al. 2004; Bresolin et al. 2005). The T_e method itself is not without difficulties, however, as temperature gradients or fluctuations in the H II regions may affect abundance determinations at high metallicities (Stasińska 2005). The result of all this is that absolute values of metal abundances are still quite uncertain; but, fortunately for our present purposes, relative abundances of similar objects, determined with the same method, are more reliable.

Our options for determining the chemical abundances of the $z \sim 2$ galaxies are limited. NIRSPEC can observe only one band in a single exposure, and at these redshifts Hα and [N II]$\lambda6584$ fall in the K-band; [O III]$\lambda\lambda5007,4959$ and Hβ in the H-band; and [O II]$\lambda3727$ in J. We have focused our observations on Hα in the K-band, and therefore the use of the R_{23} indicator would triple the required observing time. Our only option to determine the metallicity of the vast majority of the galaxies in our sample is thus the ratio of [N II] to Hα. In addition to minimizing the observing time required to obtain a large sample, this ratio has the advantages that it is insensitive to reddening and is not affected by the relative uncertainties in flux calibration of spectra taken in different bandpasses.

The use of $N2 \equiv \log [N\ II]\lambda6584/H\alpha$ as an abundance indicator was proposed by Storchi-Bergmann et al. (1994), and has been further discussed and refined by Raimann et al. (2000), Denicoló et al. (2002) and Pettini & Pagel (2004). The [N II]/Hα ratio is affected by metallicity in two ways. First, there is a tendency for the ionization parameter of an H II region to decrease with increasing metallicity (see, for example, Figure 3 of Dopita 2005), thereby increasing the ratio [N II]/[N III] and hence [N II]/Hα. Second, nitrogen has both a primary component whose abundance varies at the same rate as other primary elements, and a secondary component that increases in abundance with increasing metallicity, further raising the [N II]/Hα ratio. Using electron temperature measurements to determine ionic abundances in 20 extragalactic H II regions, Kennicutt et al. (2003) find that the nitrogen abundance is adequately described by a simple model with a primary component with constant $\log(N/O) = -0.15$ and a secondary component for which $\log(N/O) = \log(O/H) + 2.2$. In other words, secondary nitrogen becomes important for $12 + \log(O/H) \gtrsim 8.3$; unless we have significantly overestimated the metallicities of the galaxies in our sample, most of our objects are in this regime. As emphasized by Kewley & Dopita (2002), a drawback of the $N2$ indicator is its sensitivity to the ionization parameter as well as to metallicity. In addition, the [N II]/Hα ratio cannot be used to determine metallicities above approximately solar value, at which the $N2$ index saturates as nitrogen becomes the dominant coolant.

Finally, the [N II]/Hα ratio is sensitive to contamination from AGN and shock excitation, and if possible these should be ruled out before using it as a metallicity indicator. This is generally done through a diagnostic line ratio diagram such as [O III]/Hβ vs. [N II]/Hα (Baldwin et al. 1981; Veilleux & Osterbrock 1987); on such a diagram, normal galaxies and AGN fall in generally well-defined regions. Unfortunately we lack the data to place nearly all of our objects on such a diagram; because we have so far obtained very few [O III]/Hβ spectra, we have measurements of all four lines for only four galaxies. We plot [O III]/Hβ vs. [N II]/Hα for these galaxies in Figure 5.1, along with $\sim 96,000$ objects from the SDSS (small grey points), the local starbursts analyzed by Kewley et al. (2001b) (small black points), and the $z = 2.2$ galaxy (green ×) discussed by van Dokkum et al. (2005), which shows evidence of shock ionization or an AGN and which falls in a clearly different region of the diagram than our galaxies. The dashed blue line shows the maximum theoretical starburst line determined from photoionization modeling by Kewley et al. (2001a); for realistic combinations of metallicity and ionization parameter, models of normal starbursts fall below and to the left of this line. The lower blue dotted line is a similar, empirically determined classification line derived for the SDSS objects by Kauffmann et al. (2003).

Our galaxies appear to fall between the two lines, along a sequence with a higher [O III]/Hβ ratio at a given [N II]/Hα ratio (or a higher [N II]/Hα ratio for a given [O III]/Hβ ratio) compared to the SDSS galaxies.[1] A similar offset is seen in star-forming galaxies at $z \sim 1.4$ (Shapley et al. 2005a). The SDSS contains very few objects in this region, indicating that there are important physical differences between the local and high redshift samples. One plausible way to produce such a shift in the [N II]/Hα–[O III]/Hβ diagram is through some combination of a harder ionizing spectrum and an increase in electron density. Though they are weak, the density-sensitive [S II] lines in our composite spectra indicate an average electron density of $n_e \sim 500$ cm^{-3}, higher than that found in normal local galaxies but similar to densities seen in local starbursts (Kewley et al. 2001b). Furthermore, Kewley et al. (2001a) show that a relatively hard EUV radiation field is required to model the line ratios of local starbursts; note that a shift is also observed between the local starbursts and the SDSS sample. Detailed photoionization modeling would be required to determine quantitatively the origin of this shift and its effect on the calibration of the $N2$ index, and for the moment the absolute calibration of our metallicity scale remains uncertain (though the relatively shallow slope of the relation between $N2$ and (O/H) means that offsets are unlikely to be large). As diagnostic line ratios for more high redshift galaxies are measured, it is hoped that full photoionization modeling, with improved spectra of the Wolf-Rayet stars that dominate the EUV radiation field, will clarify the physical reasons at the root of the shifts in the [N II]/Hα–[O III]/Hβ diagram.

We can exclude an AGN contribution to our sample as a whole based on the X-ray and UV properties of the galaxies. Specifically, Reddy et al. (2005) find that only 3% of UV-selected galaxies at $z \sim 2$ have X-ray detections in the ultradeep 2 Ms *Chandra* Deep Field North. These galaxies, all of which have $K_s < 20$, are not included in the composite spectra considered in the present analysis. Further information on AGN contamination comes from the rest-frame UV spectra of our galaxies, which allow us to reject AGN on an individual basis in fields without deep X-ray data. As we describe further in §5.4, we have constructed six composite UV spectra, binned by stellar mass in the same way as the NIRSPEC spectra (see below), as well as larger composites of the two highest and two lowest mass bins. None of the composites show AGN features such as broad, high ionization emission lines, placing further limits on low-level AGN activity. On the basis of all of these considerations, we conclude that a significant AGN component to the galaxies considered here is unlikely, and proceed under the assumption that the emission lines we observe are produced in H II regions photoionized by hot stars.

Clearly the $N2$ method is not the ideal way to determine metallicities. However, our primary concern here is with relative, average abundances determined from composite spectra (see below), which can be determined with more accuracy than the metallicities of individual objects. We use the calibration of Pettini & Pagel (2004), which is based on H II regions whose ionic abundances have been determined from the electron temperature or from detailed photoionization modeling. From a sample of 137 such H II regions, 131 of which have abundances determined with the T_e method, Pettini & Pagel (2004) find

$$12 + \log(\mathrm{O/H}) = 8.90 + 0.57 \times N2, \tag{5.1}$$

with a 1 σ dispersion of 0.18 dex. They conclude that the $N2$ calibrator allows a determination of the oxygen abundance to within a factor of ~ 2.5 with 95% confidence, an accuracy comparable to that of the R_{23} method.

A further difficulty of the $N2$ method is that the [N II] line is weak, and is generally detected in the

---

[1]We have not corrected the Hβ fluxes for stellar absorption, but we do not expect this to be a significant effect. We expect the stellar Hβ absorption line to have an equivalent width $W_{\mathrm{abs}} \lesssim 5$ Å (Kewley et al. 2001b; Kobulnicky et al. 1999), while, assuming a typical ratio of $W_{\mathrm{H}\alpha}/W_{\mathrm{H}\beta} \sim 5$ (Leitherer et al. 1999), our galaxies have $W_{\mathrm{H}\beta} \sim 50$ Å.

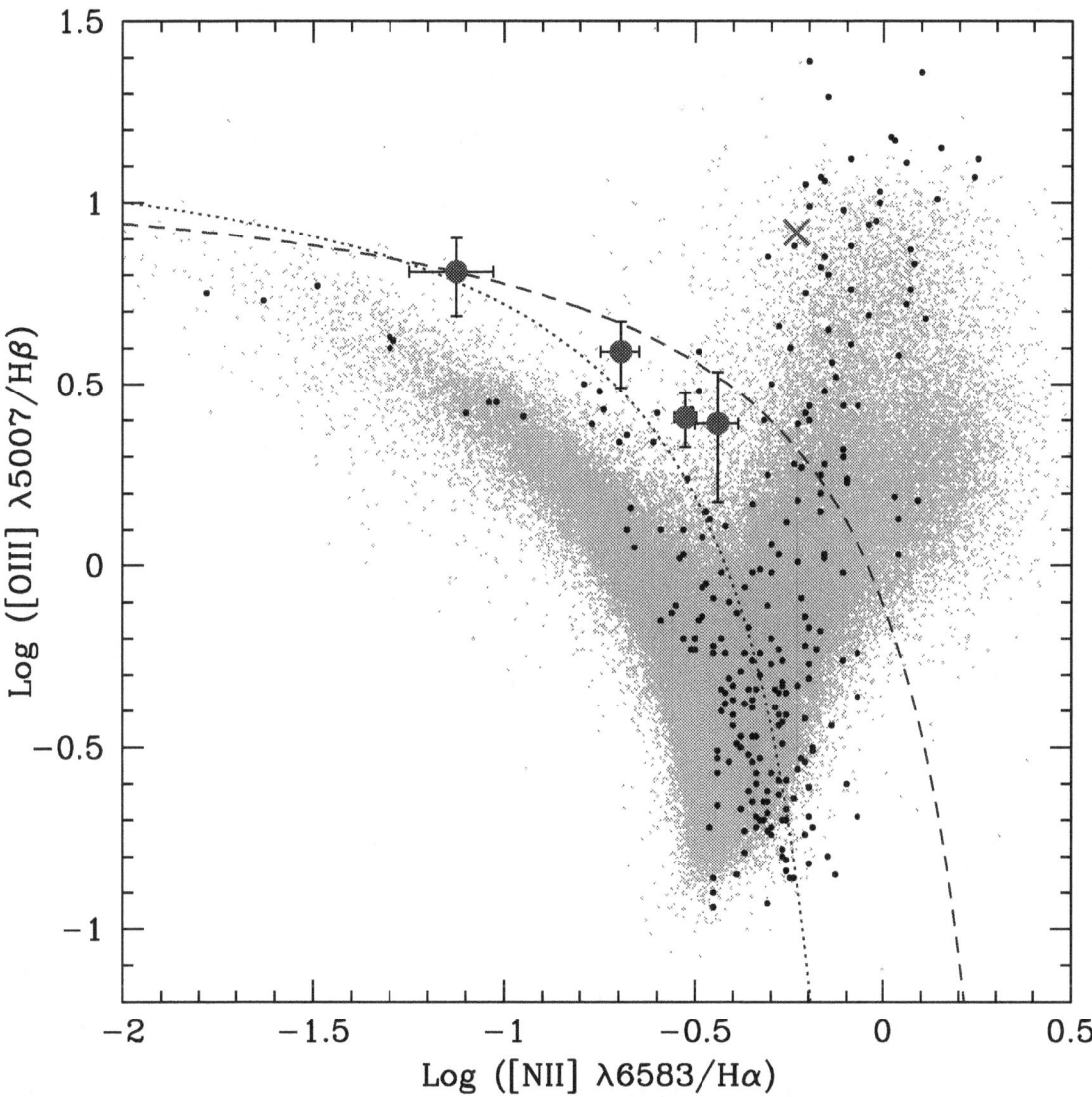

Figure 5.1 The [O III]/Hβ vs. [N II]/Hα diagnostic diagram. The four galaxies in our sample for which we have measurements of all four lines are shown by the large red circles. The $z = 2.2$ galaxy discussed by van Dokkum et al. (2005), which shows evidence of an AGN or shock ionization by a wind, is shown by the green ×. The small grey points represent ∼ 96,000 objects from the SDSS, and the small black points are the local starburst galaxies studied by Kewley et al. (2001b). The dashed line shows the maximum theoretical starburst line of Kewley et al. (2001a); for realistic combinations of metallicity and ionization parameter, star-forming galaxies fall below and to the left of this line. The dotted line is a similar, empirical determination by Kauffmann et al. (2003).

spectra of individual objects in our sample only when they approach solar metallicity (Shapley et al. 2004) or have especially strong line fluxes. In order to increase the S/N and improve the likelihood of detecting [N II] at lower metallicities, we have divided the sample into six bins by stellar mass, with 14 or 15 galaxies in each bin, and constructed a composite Hα/[N II] spectrum of the objects in each bin. The use of composite spectra has the advantage that, in addition to increasing S/N, it minimizes the effects of the mass and metallicity uncertainties of individual objects, because we are only concerned with the average properties of the galaxies in each bin. We first shift the flux-calibrated spectra into the rest-frame, and then average them, rejecting the minimum and maximum value at each point. The six composite spectra, labeled with the mean stellar mass in each bin, are shown in Figure 5.2.

We measure the [N II]/Hα ratio by first measuring the Hα fluxes, central wavelengths, and widths, and then constraining the [N II] line to have the same width and a central wavelength fixed by the position of Hα. We use the rms of the spectrum between emission lines to determine the typical noise in each spectrum; because the galaxies are at different redshifts, the systematic effects of the night sky lines are minimized in the composites, and the rms provides an adequate description of the noise. This procedure provides a good fit to the [N II] line for all of the composite spectra (except in the lowest mass bin, where we determine an upper limit on the [N II] flux). The measured Hα and [N II] fluxes for each composite, and the inferred value of 12+log(O/H), are given in Table 5.2. The listed uncertainties in 12+log(O/H) include the scatter in the $N2$ calibration (a 1σ uncertainty of 0.18 dex reduced by $\sqrt{N}$, where N is the number of objects in the composite spectrum) as well as the uncertainties in the measurements of the Hα and [N II] fluxes.

5.4   The Mass-Metallicity Relation

In Figure 5.3 we plot the mean metallicity of the galaxies in each mass bin against their mean stellar mass (large filled circles), and find a monotonic increase in metallicity with stellar mass. The vertical error bars show the uncertainty in $12 + \log(O/H)$ from measurement uncertainties in the [N II]/Hα ratio, while the additional vertical error bar in the lower right corner shows the uncertainty due to the scatter in the $N2$ calibration. The horizontal bars show the range of stellar masses in each bin. The most massive galaxies in our sample have close to solar metallicities, a result found previously by Shapley et al. (2004), who measured the [N II]/Hα ratio from individual spectra of the brightest objects with $K_s < 20$. More typical galaxies have $12 + \log(O/H) \sim 8.4$, while for the lowest-mass objects we can only place an upper limit $12 + \log(O/H) < 8.2$, or $(O/H) < 1/3 \, (O/H)_\odot$.

We have considered possible systematic effects, but have not found any that could spuriously produce the clear correlation between stellar mass and metallicity that we have found. Specifically, it is possible that we have underestimated the value of $M_\star$ in the lowest-mass bins if an older stellar population is already in place in these galaxies (see the discussion in §5.3.1). This would have the effect of steepening the observed correlation, as the lowest-mass bins would move to the right in Figure 5.3, while the high-mass bins would essentially remain unaffected. It is hard to imagine how AGN contamination could produce the correlation, given the low fraction of AGN in the sample. Variations in the ionization parameter are also unlikely to be correlated with the assembled stellar mass. The ionization parameter of an H II region depends on the age of the ionizing cluster (which is very much less than the age of the galaxy) and much less on its mass (see, e.g., Dopita 2005). Since each galaxy in our sample presumably contains many H II regions of different ages, the overall variation in ionization parameter from galaxy to galaxy should be small. We therefore have no reason to expect a dependence on the total stellar mass (other than the dependence of the ionization parameter on metallicity, which is included in the $N2$ calibration).

The dashed line in Figure 5.3 shows the mass-metallicity relation determined for $\sim 53,000$ star-forming SDSS galaxies by T04, after an arbitrary downward shift of 0.56 dex. With this shift the SDSS relation

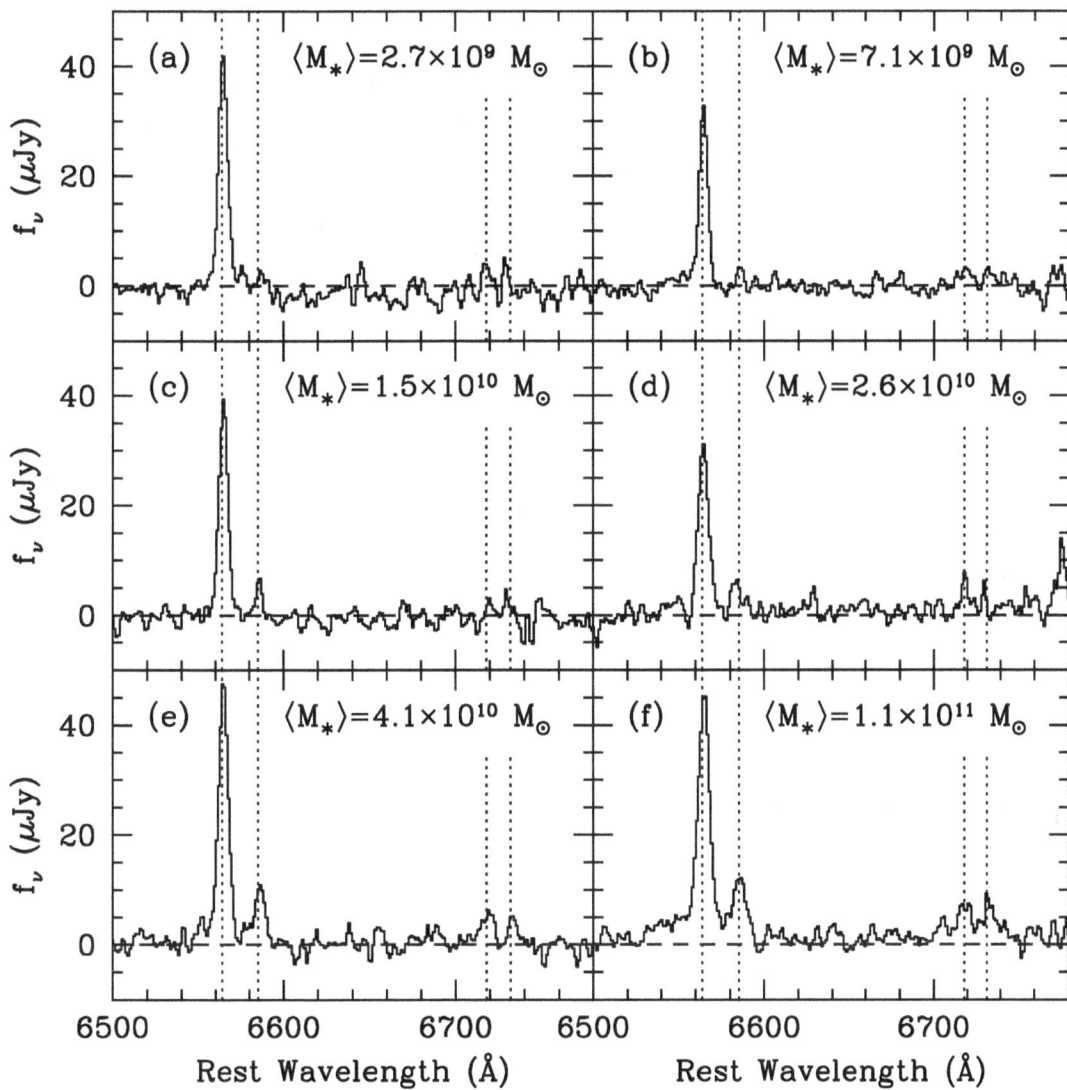

Figure 5.2 The composite NIRSPEC spectra of the 87 galaxies in our sample, divided into six bins of 14 or 15 objects each by increasing stellar mass (panels *a* through *f*). The spectra are labeled with the mean stellar mass in each bin, and the Hα, [N II], and [S II] lines are marked with dotted lines (left to right, respectively). The increase in the strength of [N II] with stellar mass can be seen clearly. The density-sensitive [S II] lines, while weak, indicate a typical electron density of $n_e \sim 500$ cm^{-3}, with no significant dependence on mass; this is a value comparable to that seen in local starburst galaxies (Kewley et al. 2001b).

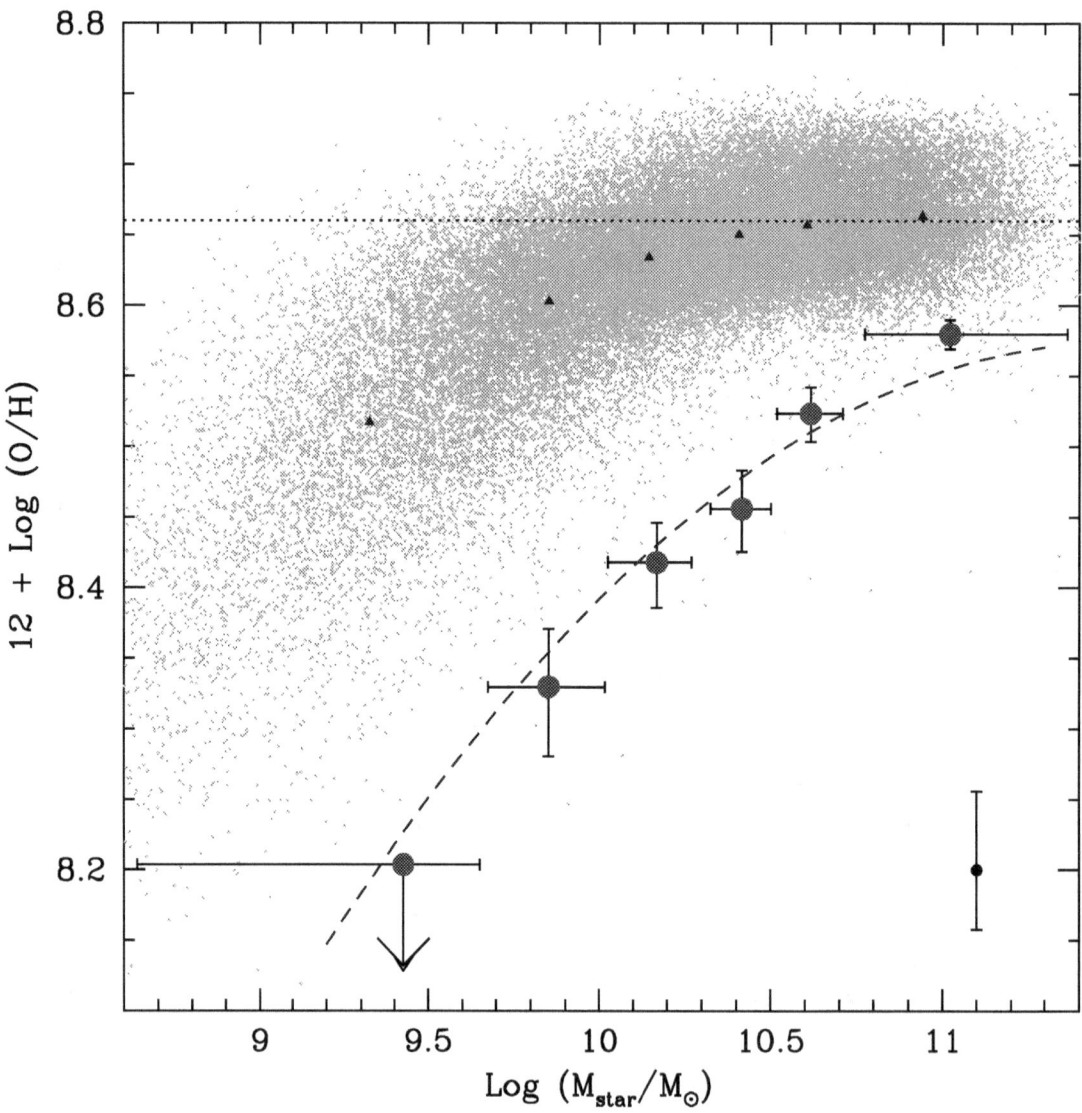

Figure 5.3 The observed relation between stellar mass and oxygen abundance at $z \sim 2$ is shown by the large red circles. Each point represents the average value of 14 or 15 galaxies, with the metallicity estimated from the [N II]/Hα ratio of their composite spectrum. Horizontal bars indicate the range of stellar masses in each bin, while the vertical error bars show the uncertainty in the [N II]/Hα ratio. The additional error bar in the lower right corner shows the additional uncertainty in the $N2$ calibration itself. The dashed blue line is the best-fit mass-metallicity relation of Tremonti et al. (2004), shifted downward by 0.56 dex. The metallicities of different samples are best compared using the same calibration; we therefore show, with small grey points, the metallicities of the $\sim 53,000$ SDSS galaxies of Tremonti et al. (2004) determined with the $N2$ index. Note that the [N II]/Hα ratio saturates near solar metallicity (the horizontal dotted line). The blue triangles indicate the mean metallicity of the SDSS galaxies in the same mass bins we use for our sample; using the more reliable, low-metallicity bins, our galaxies appear to be ~ 0.3 dex lower in metallicity at a given mass, though this may be affected by the systematics discussed in the text.

can be seen to match the $z \sim 2$ galaxies remarkably well, though it may be somewhat shallower. The empirical shift of 0.56 dex includes a possible offset due to the different abundance diagnostics used in the two studies—while the SDSS metallicity determinations take into consideration all of the strong nebular lines, ours are based on the $N2$ index alone, for the reasons explained above. For a more consistent comparison, we use the $N2$ calibration to calculate the metallicities of the same 53,000 SDSS galaxies, shown with small grey points in Figure 5.3; the mean $N2$ metallicity, in bins spanning the same range of stellar masses used for our sample, is shown by the small blue triangles. The saturation of the [N II]/Hα ratio at metallicities approaching solar (the horizontal dotted line) is clearly apparent in the SDSS sample, and makes the determination of the offset between the two samples somewhat difficult. For galaxies with $M_\star \sim 2 \times 10^9$ M$_\odot$, the oxygen abundance of the $z \sim 2$ sample is lower by > 0.3 dex; at higher stellar masses the offset is smaller, due at least in part to the saturation of the [N II]/Hα ratio. If we consider the two lowest-mass bins as the most reliable indicators, there is an apparent offset of ~ 0.3 dex in metallicity between galaxies with the same stellar mass at $z \sim 2$ and now.

There could be several reasons for this factor of ~ 2 difference. The SDSS metallicities are biased toward the innermost regions of the galaxies observed, where the spectrograph fibers were positioned. According to a recent reanalysis of this effect by Kewley & Ellison (2005), this fact alone can account for an ~ 0.15 dex offset between nuclear and global (i.e., integrated over the whole galaxy) metallicities. Some uncertainty in our metallicity scale is also indicated by the offsets between the [O III]/Hβ and [N II]/Hα ratios relative to the SDSS galaxies seen in Figure 5.1, though the effects of these offsets cannot be quantified without detailed modeling of a larger sample with all four lines. Finally, there may be a real evolutionary offset in the mass-metallicity relation; galaxies that have assembled a given stellar mass may genuinely have attained lower metallicities at earlier times (e.g., Maier et al. 2004). The relative importance of these different possibilities will no doubt be investigated in the years ahead.

In any case, it is important to remember that, while such comparisons between high and low redshift galaxies are of obvious interest, it is very likely that the two samples do *not* form an evolutionary sequence. The likely descendants of the $z \sim 2$ sample can be identified by comparing their clustering properties, evolved to $z \sim 0$, with those of objects in the local universe; such a comparison shows that early-type galaxies in the SDSS match the clustering properties of the $z \sim 2$ sample, while the later-type star-forming objects studied by T04 are too weakly clustered to be the descendants of the $z \sim 2$ population (Adelberger et al. 2005b). In this sense it is perhaps more relevant to compare the metallicities of our sample with those of local ellipticals, but because metal abundances in elliptical galaxies are determined using methods very different from ours (usually absorption features in the integrated spectra of old stellar populations), such a comparison is likely to suffer from serious systematic uncertainties. In general, though, such (non dwarf) ellipticals have near solar or super solar metallicities (e.g., Tantalo et al. 1998; Worthey 1998), consistent with those determined for at least the more massive galaxies in our sample at $z \sim 2$.

5.4.1 Composite Ultraviolet Spectra

Given the uncertainties in the absolute metallicity scale associated with the $N2$ index, it would be highly desirable to obtain independent abundance measures for the galaxies considered in this work. As discussed by Rix et al. (2004), the ultraviolet spectrum of star-forming galaxies is rich in stellar spectral features, which, in principle at least, could provide abundance diagnostics for the young stellar populations. The difficulty is that these are mostly low-contrast features, generally requiring data of higher quality than can be obtained with current instrumentation. Nevertheless, it is worthwhile examining whether the rest-frame ultraviolet spectra of the galaxies under study are consistent with the abundance trend revealed by Figure 5.3.

To this end, we have constructed two composite spectra, each consisting of approximately 30 galaxies, by averaging the LRIS-B spectra of the galaxies in, respectively, the two lower- and the two higher-mass bins in Figure 5.3. The corresponding mean stellar masses are $\langle M_\star \rangle = 5 \times 10^9$ and $\langle M_\star \rangle = 7 \times 10^{10}$ M$_\odot$. The coarser mass binning was required to improve the S/N of the rest-UV composites to the level where the stellar absorption features that are sensitive to metallicity could be clearly discerned. The two composite spectra are shown in Figure 5.4, after normalization to the underlying stellar continua following the prescription by Rix et al. (2004).

Within the spectral region covered by the composites, 1150–1925 Å, the interval near 1400 Å is particularly suitable for an abundance analysis; we show this portion on an expanded scale in Figure 5.5. The region includes two blends of stellar photospheric lines, centered at 1370 Å and 1425 Å, whose strengths were shown by Leitherer et al. (2001) to be mostly sensitive to metallicity. The "1370" feature is a blend of O V λ1371 and Fe V $\lambda\lambda$1360–1380, while Si III λ1417, C III λ1427, and Fe V λ1430 make up the "1425" feature. This portion of the spectra also encompasses the Si IV $\lambda\lambda$1393, 1402 doublet, which has a broad stellar P-Cygni component formed in the expanding atmospheres of late O and early B supergiants; superposed on this broad component are narrower interstellar absorption lines due to the ambient interstellar medium that lies in front of the stars.

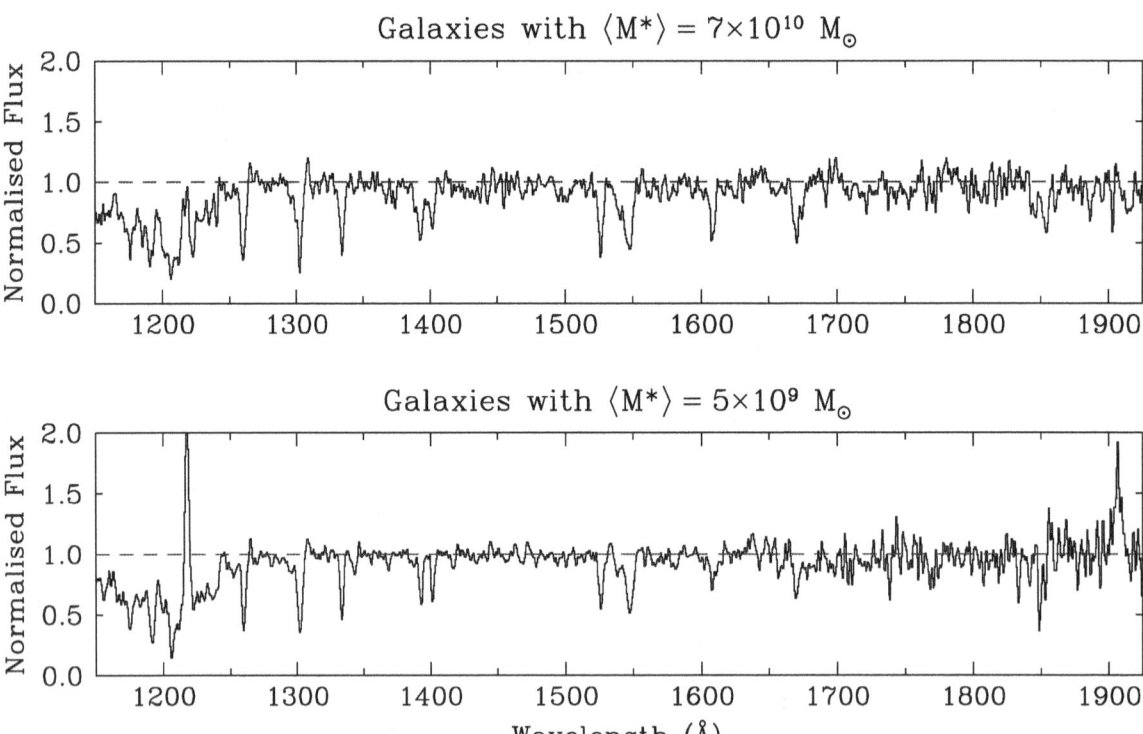

Figure 5.4 Composite rest-frame UV spectra of galaxies in the two higher (top panel) and in the two lower (bottom panel) mass bins in our sample. The upper spectrum is the average of 28 LRIS-B spectra and the mean stellar mass is $\langle M_\star \rangle = 7 \times 10^{10}$ M$_\odot$, whereas 30 LRIS-B spectra contributed to the lower composite for which $\langle M_\star \rangle = 5 \times 10^9$ M$_\odot$. The spectra have been divided by the underlying stellar continuum estimated according to the prescription by Rix et al. (2004). Differences in the stellar, interstellar, and nebular lines between the two composites are discussed in the text.

In Figure 5.5 we have also reproduced synthetic spectra generated with the Starburst99 code (Leitherer et al. 1999, 2001) for the standard case of continuous star formation and Salpeter initial mass function, using in

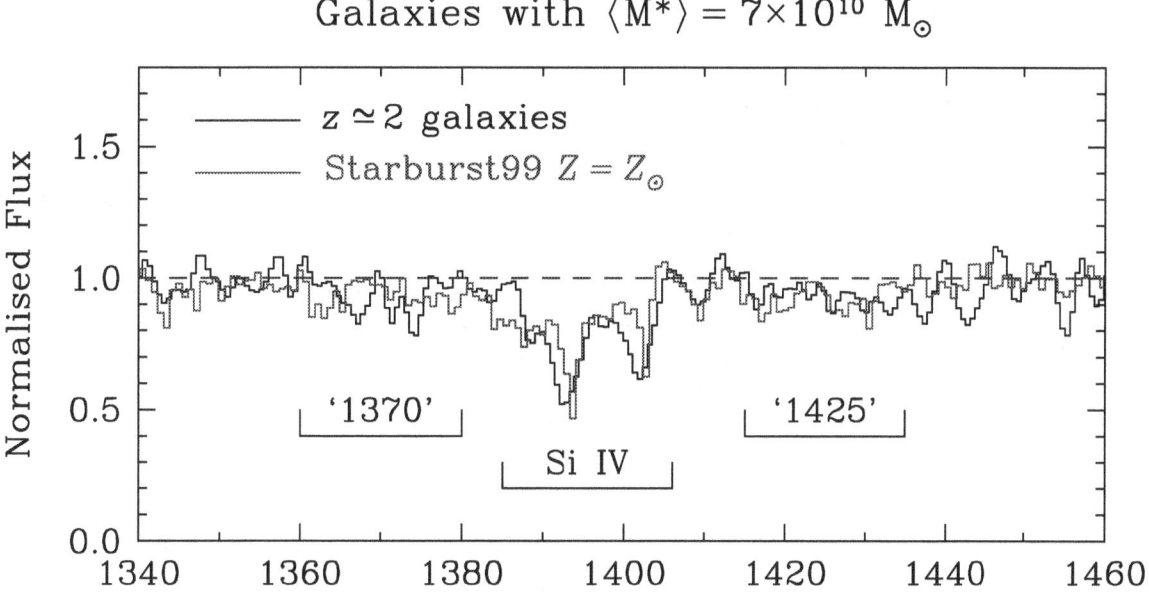

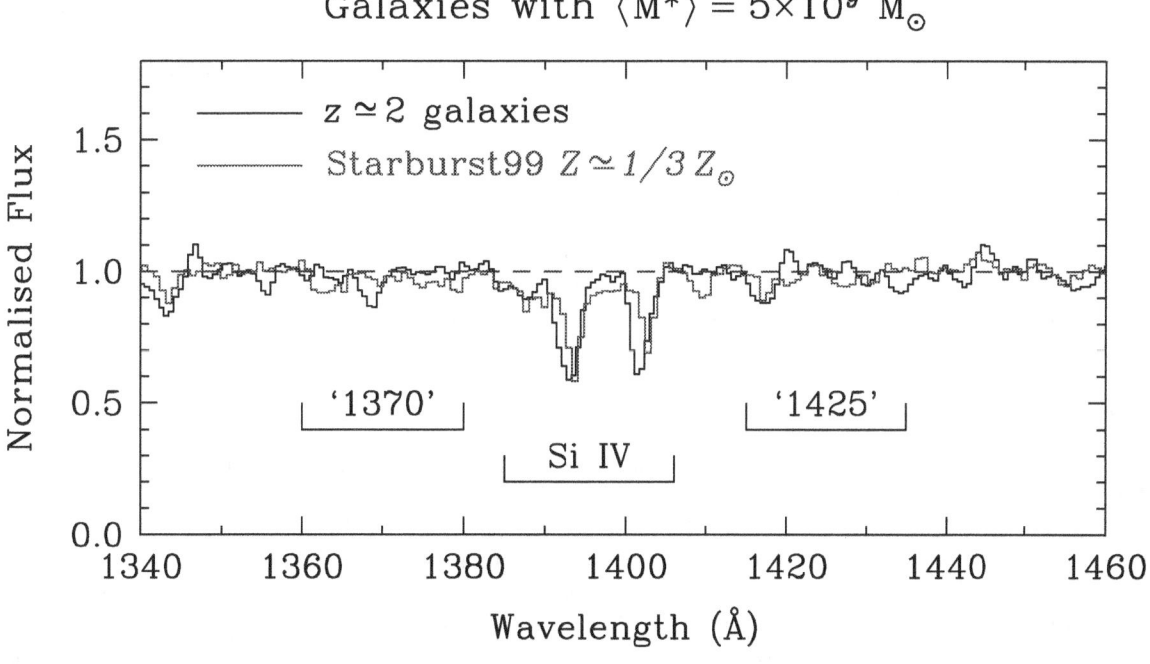

Figure 5.5 Close-up of the 1400 Å region in the two composite spectra shown in Figure 5.4 (black histogram). This region contains two blends of stellar photospheric lines, labeled "1370" and "1425," whose strength is thought to depend primarily on metallicity, and the Si IV $\lambda\lambda 1393, 1402$ doublet, which consists of a broad P-Cygni stellar absorption on which narrower interstellar lines are superposed. The red histogram shows the Starburst99 spectrum for the standard case of continuous star formation with a Salpeter IMF and, respectively, solar (top panel) and "Magellanic Cloud" (lower panel) metallicities.

turn the two empirical stellar libraries available in the code, built from spectra of OB stars in, respectively, the Milky Way (the solar metallicity, or $Z = Z_\odot$, library) and the Magellanic Clouds (the MC metallicity, or $Z = 1/3\,Z_\odot$, library). Figure 5.5 shows that the former is a plausible match to the composite spectrum of the higher stellar mass galaxies, and the latter is consistent with the composite UV spectrum of the galaxies in the lower stellar mass bins. We now discuss this comparison in more detail.

Focusing on the top panel in Figure 5.5, we find that the overall equivalent widths of the "1370" and "1425" photospheric features are in agreement between empirical and synthetic spectra. The comparison of the Si IV $\lambda\lambda 1393, 1402$ P-Cygni component is complicated by the blending with the interstellar lines, which, as is usually the case, are stronger (and blueshifted) in starburst galaxies than in the individual stars that make up the stellar libraries. Nevertheless, the broad component of the Si IV feature does appear to have comparable optical depth to the solar metallicity Starburst99 model. Turning to the lower panel in Figure 5.5, we see that all three spectral features are undetected in the lower stellar mass UV composite. The Magellanic Cloud metallicity model spectrum also shows these lines to be much reduced in strength, with the P-Cygni component of Si IV perhaps marginally stronger than observed.

The S/N ratio of the data and the subtlety of the spectral features in question limit the above comparison to qualitative statements, and we do not consider a more quantitative approach (such as a χ^2 analysis, for example) to be warranted in the present circumstances. Nevertheless, with only the UV spectra at our disposal, we would have concluded that the galaxies with a mean stellar mass $\langle M_\star \rangle = 7 \times 10^{10}$ M$_\odot$ have a metallicity $Z \sim Z_\odot$ (or $12 + \log{(O/H)} \sim 8.6$), and that the ones with $\langle M_\star \rangle = 5 \times 10^9$ M$_\odot$ have $Z \lesssim 1/3\,Z_\odot$ (or $12 + \log{(O/H)} \lesssim 8.1$). Given all the uncertainties, these conclusions are broadly consistent with those deduced from our analysis of the [N II]/Hα ratios.

Before concluding this section, we briefly point out that there are other differences between the two composite UV spectra reproduced in Figure 5.4. The interstellar absorption lines are significantly stronger in the galaxies with higher stellar mass. The strongest UV interstellar lines, Si II $\lambda 1260$, O I $\lambda 1302$+Si II $\lambda 1304$, C II $\lambda 1334$, Si II $\lambda 1526$, Fe II $\lambda 1608$, and Al II $\lambda 1670$, have rest-frame equivalent widths $W_0 = 1.5 - 2$ Å in the $\langle M^* \rangle = 5 \times 10^9$ M$_\odot$ composite, and $W_0 = 2 - 3$ Å in the $\langle M^* \rangle = 7 \times 10^{10}$ M$_\odot$ composite. Since these lines are all strongly saturated (e.g., Pettini et al. 2002), the higher equivalent widths are much more likely to be due to a larger velocity dispersion of the absorbing gas, than to an increase in the column densities of the metals. It is interesting to speculate that this may be related to differences in the star-formation histories of the two samples of galaxies; on average, star formation has been in progress for longer in galaxies with higher values of $\langle M_\star \rangle$ and presumably more kinetic energy has been deposited in their interstellar media, stirring the gas to higher velocity dispersions. Finally, it is intriguing that the lower stellar mass—and younger—galaxies have much stronger emission lines from H II regions, C III] $\lambda 1909$ and Lyα than the galaxies where star formation has been ongoing for a longer period of time.

5.4.2   The Luminosity-Metallicity Relation

Because the luminosity of a galaxy is far more easily determined than its mass, there are a great many more luminosity-metallicity (L-Z) relations than mass-metallicity relations in the literature (e.g., Skillman et al. 1989; Zaritsky et al. 1994; Garnett et al. 1997; Lamareille et al. 2004; Salzer et al. 2005, to name only a few). These correlations span 11 orders of magnitude in luminosity and 2 dex in metallicity, and are seen in galaxies of all types. Both the slope and the zeropoint of the relation shift depending on the bandpass in which the correlation is determined (Salzer et al. 2005); it is traditional to use the absolute B magnitude, but both the slope and the dispersion of the relation decrease as wavelength increases to the IR, probably due to extinction and the closer correspondence of the infrared luminosity to stellar mass. The metallicity-luminosity relation has been observed in galaxies at redshifts up to $z \sim 1$ (Kobulnicky et al. 2003; Lilly et al. 2003; Kobulnicky & Kewley 2004; Maier et al. 2004; Liang et al. 2004), at which point the shifting of most of the strong nebular lines into the IR makes spectroscopy much more difficult. Most of these studies show that the zeropoint of the L-Z relation evolves with redshift, so that galaxies of a given luminosity have decreasing metallicity with increasing redshift. The so-far small number of metallicity-luminosity comparisons at $z > 2$ seem to confirm this trend, as high redshift galaxies appear to be 2–4 mag brighter than local galaxies of comparable metallicity (Kobulnicky & Koo 2000; Pettini et al. 2001; Shapley et al. 2004).

We construct a luminosity-metallicity relation analogous to our mass-metallicity relation by dividing our sample of 87 galaxies into six bins by rest-frame absolute B magnitude M_B, which we determine by multiplying the best-fit SED of each object by the redshifted B-band transmission curve. We construct a composite spectrum of the galaxies in each bin, and measure the [N II]/Hα ratio and determine the oxygen abundance in the manner described in §5.3.2. We plot the results in Figure 5.6, again including SDSS metallicities determined from the $N2$ indicator. It is immediately apparent that the correlation between luminosity and metallicity is weaker than that between mass and metallicity; the trend with luminosity is not monotonic, and is not statistically significant. A Spearman correlation test finds a 33% probability that the points are uncorrelated, giving a significance of 1σ. The faintest galaxies do appear to have lower metallicity, however; any appearance of correlation is driven by this bin.

The comparison with the SDSS sample is again made difficult by the saturation of the [N II]/Hα ratio around solar metallicity, but it is clear that the high redshift galaxies have both lower metallicities (subject to the caveats discussed in §5.4) and higher luminosities than most of the local sample. Considering the offset of the more reliable lower metallicity bins, we see from Figure 5.6 that the $z \sim 2$ galaxies are approximately 3 magnitudes brighter than local galaxies with the same oxygen abundance. It is difficult to invert the comparison to determine the difference in metallicity between galaxies of a given luminosity between the two samples, however, since virtually all of the local galaxies as bright as the $z \sim 2$ sample have solar or greater abundances that cannot be accurately determined by the $N2$ method. In order for the best-fit L-Z relation determined by T04 to pass through the average luminosity and metallicity of our sample, it must be shifted downward by ~ 0.9 dex. After allowing for the ~ 0.25 dex systematic difference between the metallicity diagnostics used by T04 and here (as discussed in §5.4), we are still left with a shift of 0.6–0.7 dex between the local and high redshift samples. Our mean values are offset from other local L-Z relations by amounts ranging from ~ 0.4 (Skillman et al. 1989) to ~ 1.0 (Lamareille et al. 2004) dex. The large offsets between the various local relations are due to differences in calibration methods and sample selection; the relations are not directly comparable, and it is not yet clear which of them, if any, provides the most appropriate comparison for our sample. While it is probably a robust conclusion that the $z \sim 2$ galaxies are more metal-poor than their luminous local counterparts, a better understanding of systematic effects is required to reliably quantify the differences.

Our results are consistent with previous L-Z determinations at $z > 2$, and, moreover, they are not surprising given our knowledge of the high redshift sample. Star-forming galaxies at $z \sim 2$ have lower average mass-to-light ratios M/L than local galaxies, and the variation in M/L at a given rest-frame optical luminosity can be up to a factor of 70 (Shapley et al. 2005b). If the fundamental correlation is between metallicity and mass, these differences would shift the relation to higher luminosities and dramatically increase the scatter, as we have observed. For this reason, a mass-metallicity correlation is more physically meaningful than a luminosity-metallicity correlation at high redshifts. While it is likely that an L-Z relation determined in the rest-frame IR would show a stronger correlation, this possibility cannot be investigated until a larger sample of high-z galaxies with both mid-IR photometry and nebular line spectra has been assembled.

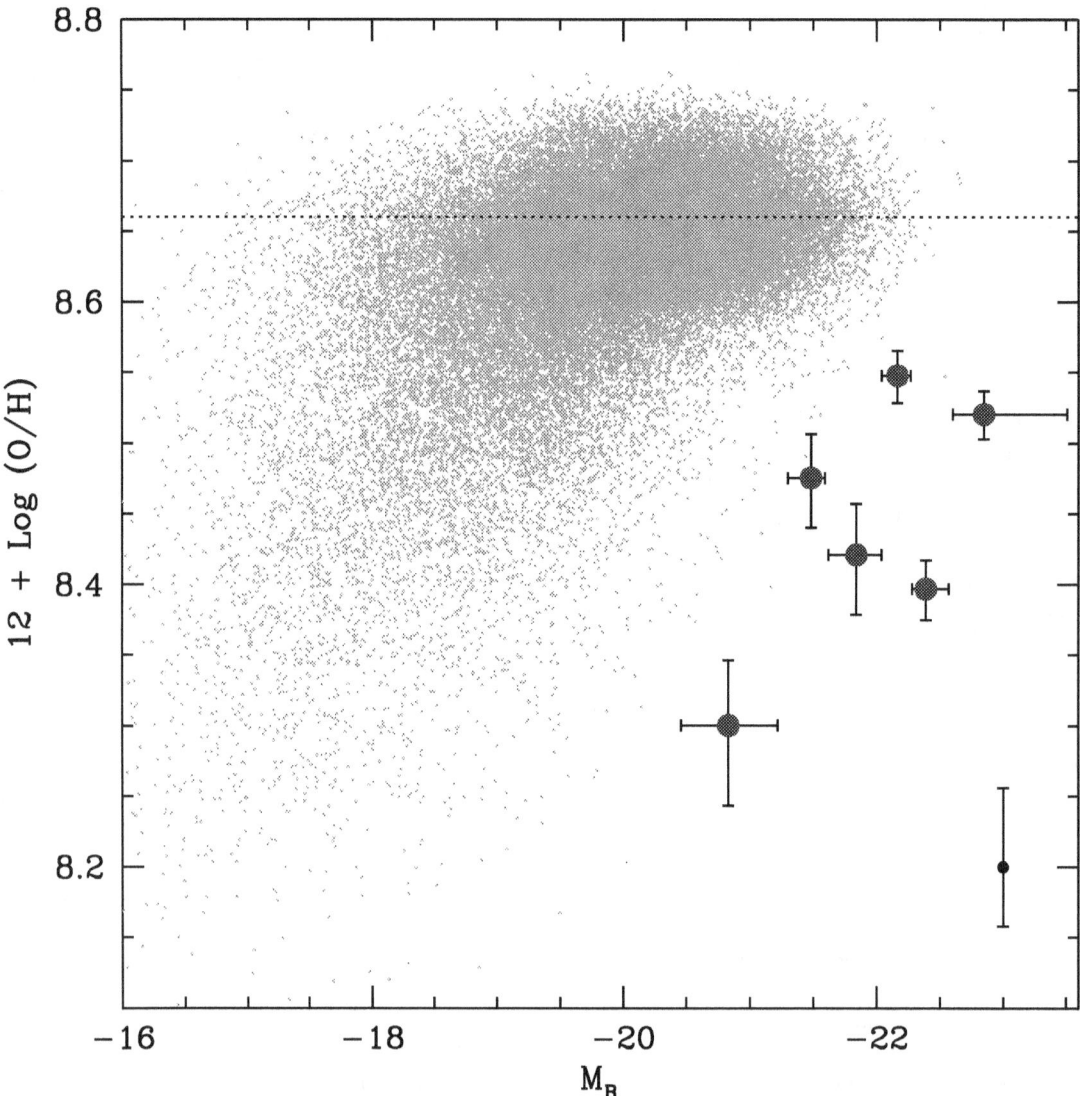

Figure 5.6 The luminosity-metallicity relation at $z \gtrsim 2$. We have divided the sample into six bins by rest-frame absolute B magnitude, and estimated the metallicity in each bin. The symbols are the same as in Figure 5.3. The points are not significantly correlated, though the faintest galaxies do appear to have lower metallicity. The lack of correlation can be understood through the large variation in the optical mass-to-light ratio at high redshift. The $z \sim 2$ galaxies are ~ 3 magnitudes brighter than the SDSS galaxies at a given metallicity, as estimated by the [N II]/Hα ratio.

5.5   The Origin of the Mass-Metallicity Relation

A correlation between gas-phase metallicity and stellar mass can plausibly be explained either by the tendency of lower-mass galaxies to have larger gas fractions (McGaugh & de Blok 1997; Bell & de Jong 2000) and thus be less enriched, or by the preferential loss of metals from galaxies with shallow potential wells by galactic-scale winds. With the relevant information on the star, gas, and metal content of the galaxies, the two effects can be differentiated. In the simple, closed-box model of chemical evolution with no inflows or outflows, the mass fraction of metals Z is a simple function of the gas fraction $\mu \equiv M_{\rm gas}/(M_{\rm gas} + M_\star)$ and the true yield y, which represents the ratio of the mass of metals produced and ejected by star formation to the mass of metals locked in long-lived stars and remnants. The true yield is a function of stellar nucleosynthesis, and as in previous, similar studies (T04, Garnett 2002) we assume that it is constant (see Garnett 2002 for a discussion). The metallicity is then given by

$$Z = y \ln(1/\mu). \qquad (5.2)$$

This equation can be inverted to determine the effective yield $y_{\rm eff}$ from the observed metallicity and gas fraction, $y_{\rm eff} = Z/\ln(1/\mu)$. If the simple model applies, $y_{\rm eff}$ will be constant for all masses and equal to the true yield y, while a decrease in $y_{\rm eff}$ (either with respect to the expected true yield or, more commonly, as a function of mass) is a signature of outflows or of dilution by the infall of metal-poor gas.

   T04 determined the effective yields of the SDSS galaxies by using the empirical Schmidt law (Kennicutt 1998b), which relates the star formation rate per unit area to the gas surface density, to estimate gas masses. They found lower effective yields in galaxies with lower baryonic masses (the baryonic mass is expected to correlate with the dark matter content, and thus indicate the depth of the potential well; McGaugh et al. 2000), and interpreted this result as evidence for the preferential loss of metals in low-mass galaxies. We carry out a similar analysis on our sample of $z \sim 2$ galaxies. As described by Erb et al. (2005b), we use a galaxy's Hα luminosity and the spatial extent of its Hα emission $r_{\rm H\alpha}$ (after deconvolution with the seeing point spread function) to estimate each galaxy's gas surface density $\Sigma_{\rm gas}$ and gas mass $M_{\rm gas} \sim \Sigma_{\rm gas} r_{\rm H\alpha}^2$. We then combine $M_{\rm gas}$ and $M_\star$ to obtain an estimate of the gas fraction μ, and compute the mean value of μ in each mass bin. Although the considerable uncertainties in both the corrected Hα flux and the galaxy's size translate into a significant uncertainty in μ for individual objects, the number of galaxies in each bin allows us to determine the mean value of μ to within $\sim 10\%$.

   As in the local universe (McGaugh & de Blok 1997; Bell & de Jong 2000), we see a strong trend of decreasing gas fraction with increasing stellar mass (and age; see Table 5.1). The lowest-mass bin in our sample has a mean gas fraction $\langle \mu \rangle = 0.85$, while the highest-mass bin has $\langle \mu \rangle = 0.22$. The median value of μ in our sample is ~ 0.5, significantly higher than the corresponding median value of ~ 0.2 for the SDSS galaxies considered by T04. A consequence of the large gas fractions of the low stellar mass bins is that the baryonic mass $M_{\rm gas} + M_\star$ of the galaxies in our sample spans a much smaller range than the stellar mass. The difference in mean stellar mass between the highest- and lowest-mass bins is a factor of 39, while the difference in mean baryonic mass between the same bins amounts to a factor of only six. This relatively small range in baryonic mass also limits our ability to detect the loss of metals in low-mass galaxies. We show the variations of $M_\star$, $M_{\rm gas}$, and $M_{\rm bar}$ with metallicity in Figure 5.7. Notably, the increase in baryonic mass along our sequence of six bins is driven almost entirely by the increase in stellar mass, while the gas mass remains relatively constant (the absence of galaxies with both low stellar masses and low gas masses is probably a selection effect, as such objects would likely be too faint to be detected in our K-band images). We also note that the dynamical masses derived from the Hα line widths are a better match to the baryonic masses than the stellar masses, with $M_{\rm dyn} \gg M_\star$ for the objects in the lowest mass bin. Erb et al. (2005b) discuss this comparison in detail.

   We plot the metallicity Z against the mean gas fraction μ in each mass bin in Figure 5.8, decreasing the

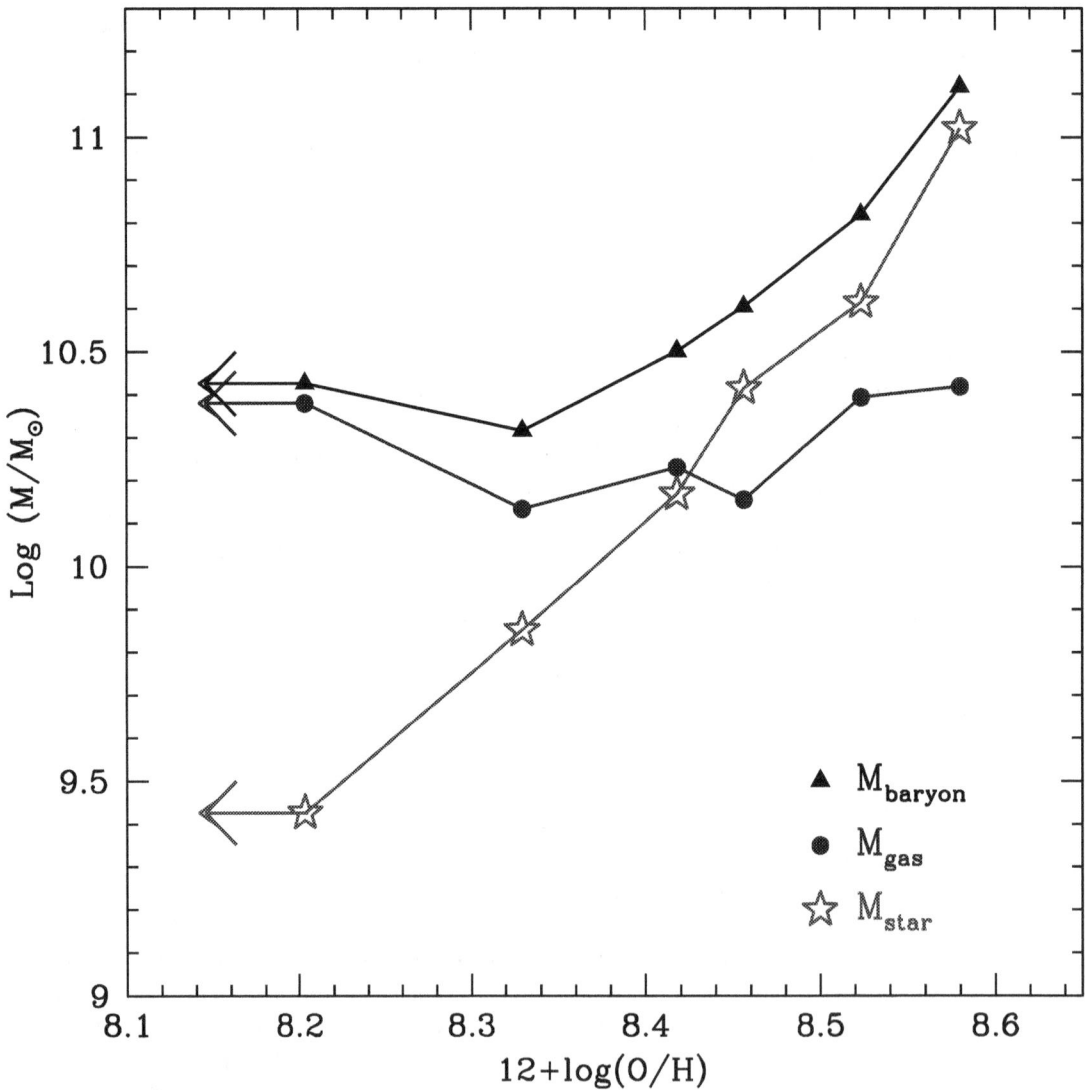

Figure 5.7 The variation of stellar, gas, and baryonic mass in each of our six bins with metallicity. Across our observed range in oxygen abundance, stellar mass increases strongly, baryonic mass increases weakly, and gas mass remains approximately constant. We thus see an increase in metallicity with decreasing gas fraction, as predicted by the closed-box model.

gas fraction from left to right to show the increase in Z as μ declines. On such a plot lines of constant yield are curves given by Eq. 5.2; the solid line shows the weighted mean of our sample, $y_{\text{eff}} = 0.008$, while the shaded region, corresponding to a range of $y_{\text{eff}} = 0.0064$ to $y_{\text{eff}} = 0.0088$, falls within the 2σ uncertainties in y_{eff} for all of the bins (we do not consider the lowest-mass bin, for which we find only an upper limit on Z and y_{eff}). The mean gas fractions and effective yields for each mass bin are given in Table 5.2. In contrast to the SDSS sample, we see no decline in the effective yield with decreasing baryonic mass; the value we find for y_{eff} in each bin is within 1σ of the weighted mean $y_{\text{eff}} = 0.008$ in 4 of our bins, and within 2σ of this value in the fifth. Our data are therefore consistent with a constant effective yield.

For the range of baryonic masses spanned by the six bins in our sample (2.1×10^{10} M$_\odot$ to 1.3×10^{11} M$_\odot$), T04 find the slope of y_{eff} as a function of baryonic mass to be shallow, with a difference of only ~ 0.0013 in y_{eff} between the two extremes in baryonic mass. This difference is comparable to our uncertainties in metallicity, and our failure to detect the signature of metal loss is not therefore evidence that it does not exist. Significant metal loss from winds occurs primarily in low-mass galaxies that are not present in our sample, however. T04 find that a galaxy with baryonic mass $M_{\text{bar}} = 3.7 \times 10^9$ M$_\odot$ loses half its metals to winds; according to their parameterization, the lowest-mass galaxies in our sample would be expected to retain $\sim 70\%$ of their metals. In any case, the fact that we observe a strong trend in metallicity over such a narrow range in baryonic mass, when wind-powered ouflows might be expected to have relatively little differential effect, suggests that the simple, closed-box picture provides a reasonable explanation of the variation of metallicity with stellar mass in the $z \sim 2$ galaxies. The salient points of this model are a constant effective yield (which is the true yield if the simple model applies), an increasing gas fraction with decreasing stellar mass, and an increase in metallicity with a decrease in gas fraction; within the uncertainties and subject to the assumptions we have made, our observations satisfy all of these conditions.

It is worth examining this result in more detail, given the strong evidence for outflows in local galaxies, and, more importantly, the ubiquitous signature of galactic-scale winds in the kinematics of the $z \sim 2$ galaxies themselves. Given sufficient range in the depth of the potential well, would we expect to be able to discern the impact of the winds on the chemical evolution of high redshift galaxies? To address this question we modify the simple model to include gas outflow at a rate of $\dot{M}$ M$_\odot$ yr^{-1}, which is a fraction f of the star formation rate. We assume that the metallicity of the outflowing gas is the same as the metallicity of the gas that remains in the H II regions in the galaxy. It can then be shown that the metallicity is given by

$$Z = y\,(1+f)^{-1} \ln\left[1 + (1+f)(\mu^{-1} - 1)\right]. \tag{5.3}$$

We show the evolution of Z with gas fraction for $f = 0$, 0.5, 1, and 2 in Figure 5.9. At high gas fractions and with $\dot{M} \sim \text{SFR}$, Z is virtually indistinguishable from the metallicity given by Eq. 5.2. We determine the gas fraction at which the two metallicities diverge by more than 3 times our typical uncertainty in Z; for a mass outflow rate equal to the star formation rate (see, e.g., Pettini et al. 2000; Martin 2003; Murray et al. 2005) this does not occur until $\mu \sim 0.3$ (the value of μ ranges from 0.2 to 0.4 for outflow rates between 0.5 and 2 times the SFR). Assuming that there is no inflow, the evolution of the gas fraction with time is a function of the star formation rate, the mass outflow rate $\dot{M}$, and the initial gas mass M_i (which is also the total initial mass). We define a timescale $\tau = M_i/\text{SFR}$; for $\dot{M} = \text{SFR}$, the gas will be entirely depleted at $t = 0.5\tau$, and half will remain at $t = \tau/3$. The gas fraction of interest, $\mu \sim 0.3$, is reached at $t \sim 0.4\tau$. The actual value of τ is quite uncertain, of course; for $M_i = 10^{11}$ M$_\odot$ and an SFR of 30 M$_\odot$ yr^{-1} (a typical value for the galaxies in the current sample; see Erb et al. 2005b), $\tau = 3$ Gyr, and $\mu = 0.3$ is reached at $t \sim 1$ Gyr. For the same SFR and $M_i = 2 \times 10^{10}$ M$_\odot$, our lowest value of M_{baryon}, $\mu = 0.3$ is reached at $t \sim 300$ Myr. Though these are order-of-magnitude values at best, they are comparable to the best-fit ages of the galaxies in our sample, suggesting that a substantial amount of time must elapse before metallicity depletion through winds could be detected with significance.

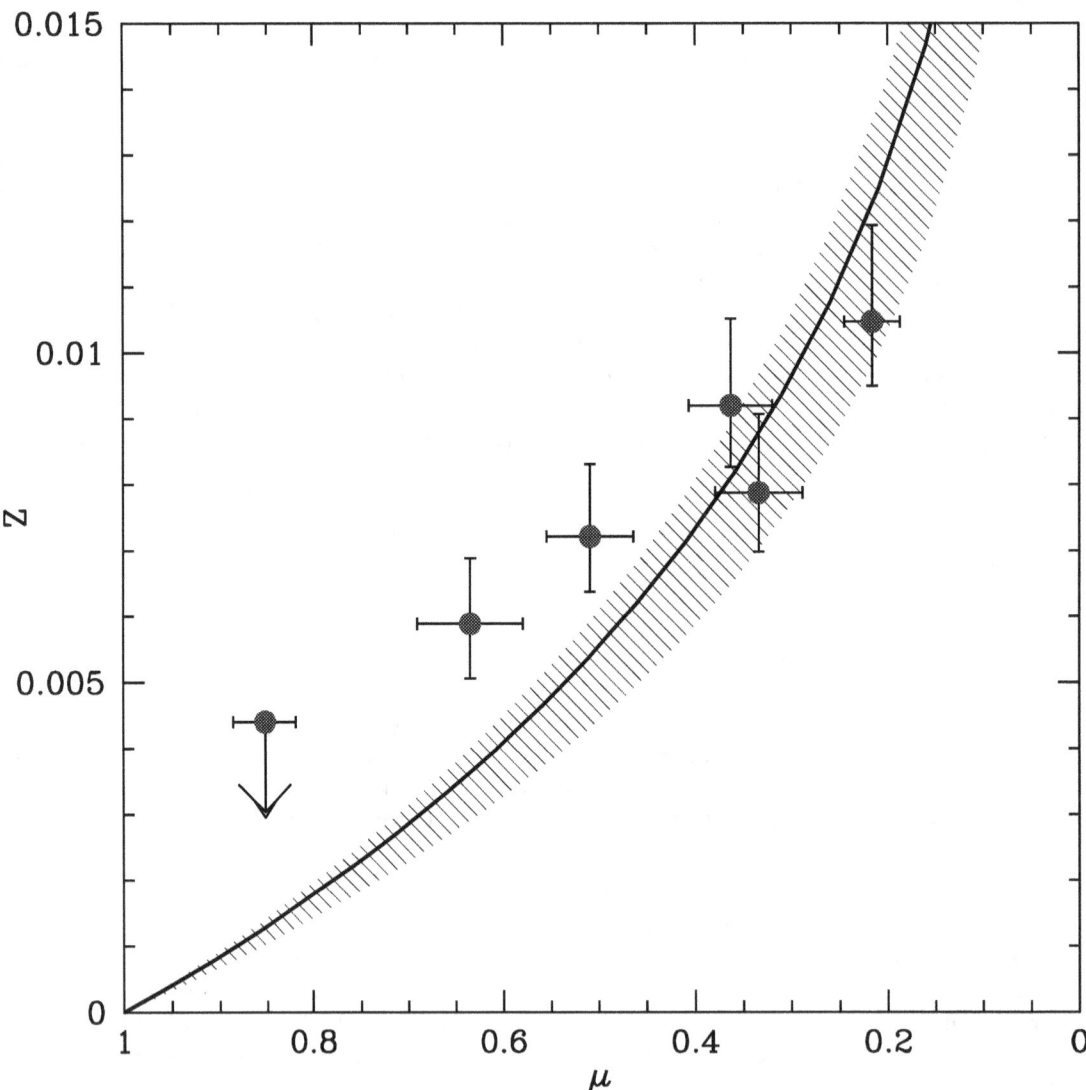

Figure 5.8 The mean metallicity Z in each mass bin, plotted against the mean gas fraction μ. Gas fraction decreases from left to right, to show the increase in metallicity with decreasing gas fraction. The solid line shows our weighted mean effective yield $y_{\mathrm{eff}} = 0.008$, while the shaded region is within the $2\,\sigma$ uncertainties in y_{eff} for all bins (we neglect the lowest stellar mass bin, with only an upper limit on Z). Within the uncertainties, our observations are consistent with a constant effective yield.

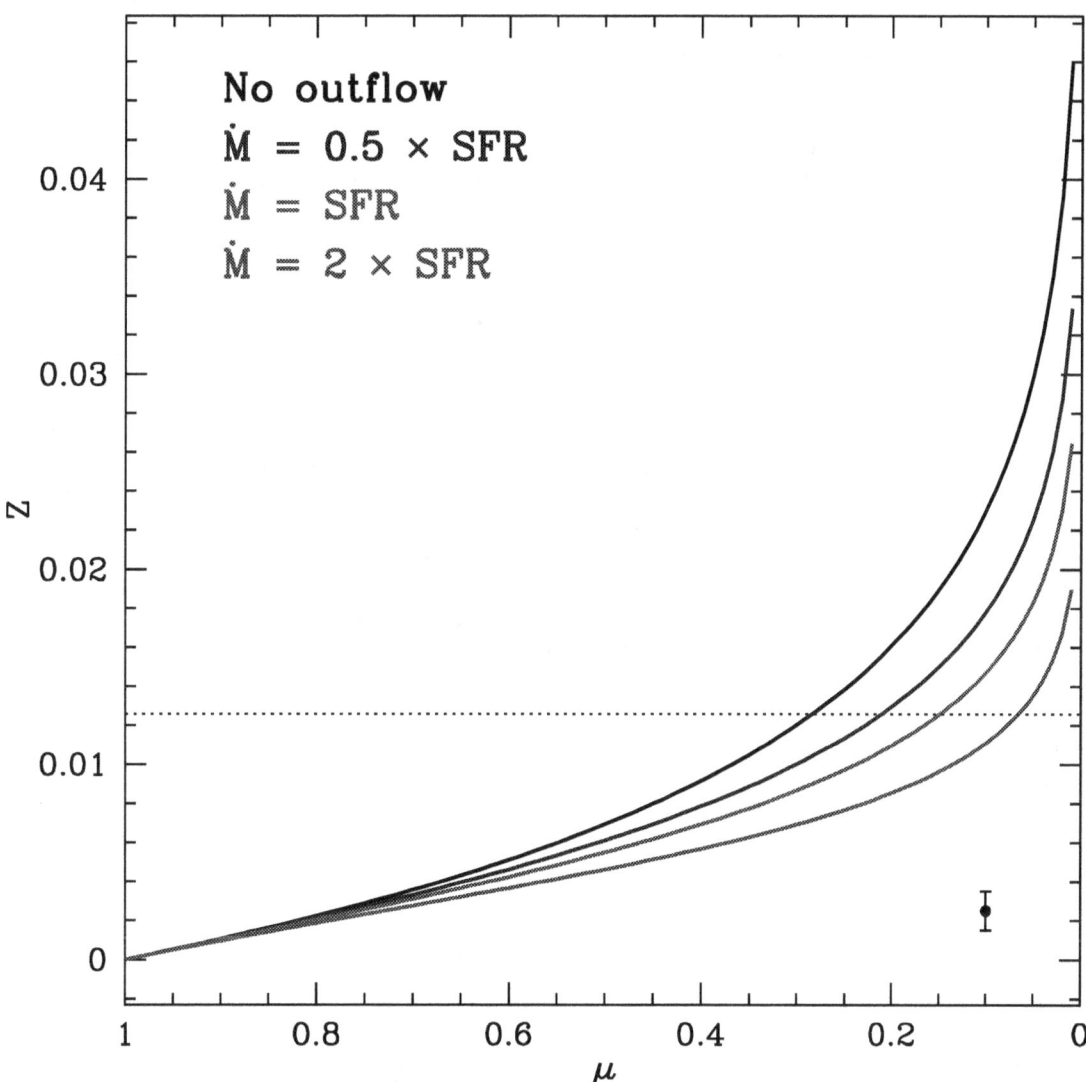

Figure 5.9 The increase of the metallicity Z with decreasing gas fraction μ for the closed-box model with no inflows or outflows (top line) and for outflow rates of 0.5, 1, and 2 times the SFR (from second to top to the bottom, blue, green, and red lines, respectively). Our typical uncertainty in Z is shown by the error bar in the lower left corner. The difference in metallicity between the simple model and the model with an outflow rate $\dot{M} = \mathrm{SFR}$ reaches three times our typical error at $\mu \sim 0.3$.

The assumption that the metallicity of the outflowing gas is the same as the observed abundances in the galaxy may not be correct. Metal-enhanced hot winds have been observed in X-rays in local starbursts; in dwarf galaxies, such winds may carry away nearly all the oxygen produced by the burst (Martin et al. 2002). Such a metal-enhanced wind would lessen the time needed to detect depletion, and may also lessen the dependence of the observed metallicity on the depth of the potential while the wind is occurring, since metals in such a hot wind will not be detected whether or not the wind will ultimately escape the galaxy. On the other hand, mergers or accretion events, which are expected to be relatively common at high redshift, would add to the gas mass and increase the time needed for noticeable depletion (unless, of course, the merger significantly increases the star formation rate). The immediate effect of the winds on the galaxies' metallicities remains uncertain, but it is unlikely that they would produce the strong observed trend in the oxygen abundance over a range of only a factor of ~ 6 in baryonic mass. Our results, combined with those of T04, therefore suggest a possibly dual origin for the mass-metallicity relation, as differences in metallicity related to the gas fraction in young galaxies are reinforced by the effects of supernova feedback with the progress of time.[2]

Substantial uncertainties remain, of course. Our derivation of the gas masses and gas fractions is indirect, and assumes that the Schmidt law takes the same form at $z \sim 2$ as in the local universe. This has not yet been tested, although the one similar galaxy with a direct measurement of the gas mass, the lensed $z = 2.7$ LBG MS1512-cB58, appears to be consistent with the local Schmidt law (Baker et al. 2004). As long as some form of the Schmidt law applies, our results will be qualitatively the same. A related question concerns the appropriateness of the gas masses derived from the Schmidt law. These represent only the gas associated with current star formation, and are therefore almost certainly an underestimate of the total gas masses. In a typical disk galaxy today, $\sim 40\%$ of the gas mass is not included by the Schmidt law (Martin & Kennicutt 2001); this fraction could plausibly be higher in the young starbursts in our sample. It is not clear how or if this gas affects the metal enrichment and star formation. Given our lack of information on this gas, we do not consider it. T04 discussed these questions of gas masses derived from the Schmidt law in somewhat more detail, but arrived at similar conclusions.

We also consider the possibility, discussed in §5.3.1 and §5.4, that we have underestimated the stellar masses of the galaxies with the smallest $M_\star$ by a factor of a few. This would primarily affect the objects in the lowest-mass bin, and would result in a decrease in the gas fraction and a decrease in the upper limit we place on the effective yield. As the current upper limit is well above the typical values we find for the effective yield in the other bins, the consistency with the simple model would be unchanged. We also find a somewhat high value of y_{eff} in the second-lowest-mass bin, such that a decrease in the effective yield would improve the agreement with the closed box model.

Finally, it is important to confirm our trend in metallicity with additional measurements and with abundance indicators that use a broader set of emission lines. The absolute values of the abundances in our sample are quite uncertain, and the improved understanding of the galaxies' physical conditions that will result from a larger sample of lines will be essential for determining this absolute scale.

5.6   Summary and Conclusions

We have used composite $\mathrm{H}\alpha/[\mathrm{N\ II}]$ spectra of 87 star-forming galaxies at $z \gtrsim 2$, divided into six bins by stellar mass, to study the correlation between stellar mass and metallicity at high redshift. Our conclusions are summarized as follows:

---

[2]In this context it is worth again pointing out that the likely descendants of the $z \sim 2$ galaxies are found among early-type galaxies at $z \sim 1$, and among ellipticals in the local universe (Adelberger et al. 2005b). By $z \sim 0$ our galaxies have lost most of their gas, whether to star formation or outflows, and so are not the objects studied by T04. Nevertheless, starburst galaxies with strong winds are the most natural progenitors of the T04 sample.

1. There is a strong correlation between stellar mass and metallicity at $z \gtrsim 2$, as the quantity $12 + \log(\text{O/H})$ increases monotonically from < 8.2 for galaxies with $\langle M_\star \rangle = 2.7 \times 10^9$ M$_\odot$ to 8.6 for galaxies with $\langle M_\star \rangle = 1.0 \times 10^{11}$ M$_\odot$. The absolute values of the oxygen abundances are uncertain, but the trend is unlikely to be due to systematic effects such as AGN contamination or variations in the ionization parameter of the H II regions.

2. Rest-frame B-band luminosity and metallicity are not significantly correlated. The $z \sim 2$ galaxies are systematically brighter than local star-forming galaxies spanning the same range in stellar mass, showing that they have smaller mass-to-light ratios M/L. At a given metallicity, local galaxies are ~ 3 magnitudes fainter than the $z \sim 2$ sample. The known large scatter in the rest-frame optical M/L at $z \sim 2$ accounts for the lack of correlation between luminosity and metallicity, and indicates that the correlation with stellar mass is (as expected) more fundamental.

3. We use the Schmidt law to estimate the gas masses and gas fractions of the $z \sim 2$ galaxies, finding that the gas fraction increases substantially with decreasing stellar mass. The lowest-mass bin in our sample has a mean gas fraction of 85%, while the highest stellar mass bin has a fraction of 20%. Our median gas fraction is $\sim 50\%$, as compared to $\sim 20\%$ in local star-forming galaxies. A consequence of the trend in gas fraction with stellar mass is a much smaller range in baryonic mass (a factor of 6) than stellar mass (a factor of 39) across the sample.

4. We use the observed metallicities and gas fractions to estimate the effective yield in each mass bin, and find that, within the uncertainties, the yield is constant in all bins at $y_{\text{eff}} \sim 0.008 = 0.6$ Z$_\odot$. The mass-metallicity trend in our galaxies thus appears to be adequately explained by the simple, closed-box model of chemical evolution. Nevertheless, the galactic winds that drive metals into the IGM are almost certainly present in our sample. We show that the range in baryonic mass considered here is too small to be able to detect the differential effects of winds on the effective yield. Moreover, the time needed for the winds to have a detectable effect on the metallicity of the galaxies is probably comparable to the ages of the galaxies themselves. We therefore suggest a dual origin for the mass-metallicity relation, as changes in metallicity naturally caused by the decrease in gas fraction as stars form are reinforced by starburst-driven winds that drive the metals out of galaxies with small potential wells.

Much remains to be done in order to improve upon the substantial uncertainties inherent in the present work. Independent measurements of the gas masses of galaxies at high redshift, though very difficult, are essential to determine whether our derived gas fractions and effective yields are valid. Additional metallicity measurements, based on other indicators that use a wider set of emission lines, are needed to confirm the trend with stellar mass revealed by the $N2$ index, to provide a better understanding of the physical conditions in the H II regions, and to establish a secure absolute calibration of the metallicity scale. More sensitive determinations of metallicity may also allow us to detect metal depletion by winds at high redshift, especially when larger telescopes bring within reach fainter galaxies, with smaller potential wells, than we are able to study at present. We anticipate that all of these measurements will greatly increase our understanding of the interplay between stars and gas, within and outside of galaxies, at high redshift.

We thank Mike Dopita and Lisa Kewley for illuminating discussions, and the staffs of the Keck and Palomar observatories for their assistance with the observations. CCS, DKE, and NAR have been supported by grant AST03-07263 from the U.S. National Science Foundation and by the David and Lucile Packard Foundation. AES acknowledges support from the Miller Institute for Basic Research in Science, and KLA from the Carnegie Institution of Washington.

Table 5.1.   Mean Stellar Masses and Stellar Population Properties

Bin	N[a]	$\langle z \rangle$[b]	Stellar Mass[c] (10^{10} M$_\odot$)	SFR[d] (M$_\odot$ yr^{-1})	Age[e] (Myr)	E(B-V)[f]
1	15	2.36	0.27 ± 0.15	55 ± 52	180 ± 276	0.19 ± 0.09
2	15	2.27	0.71 ± 0.17	24 ± 10	448 ± 374	0.14 ± 0.07
3	15	2.21	1.5 ± 0.3	30 ± 21	968 ± 944	0.11 ± 0.08
4	14	2.28	2.6 ± 0.4	28 ± 17	1026 ± 864	0.16 ± 0.07
5	14	2.21	4.1 ± 0.6	47 ± 30	1311 ± 790	0.17 ± 0.09
6	14	2.26	10.5 ± 5.4	47 ± 24	2409 ± 591	0.19 ± 0.06

[a]Number of galaxies in bin

[b]Mean $z_{H\alpha}$.

[c]Mean and standard deviation of stellar mass from SED fitting; we use a Chabrier (2003) IMF.

[d]Mean and standard deviation of SFR from extinction-corrected Hα luminosity, including a factor of 2 aperture correction determined from narrow-band imaging and comparison of the K-band continuum with broad-band magnitudes.

[e]Mean and standard deviation of best-fit age from SED fitting.

[f]Mean and standard deviation of best-fit $E(B - V)$ from SED fitting.

Table 5.2. Oxygen Abundances and Gas Fractions

Bin	Stellar Mass[a] $(10^{10}\ M_\odot)$	$F_{H\alpha}$[b] $(10^{-17}\ \mathrm{erg\ s^{-1}\ cm^{-2}})$	$F_{[\mathrm{NII}]}$[b] $(10^{-17}\ \mathrm{erg\ s^{-1}\ cm^{-2}})$	$N2$[c]	$12 + \log(\mathrm{O/H})$[d]	M_{bar}[e] $(10^{10}\ M_\odot)$	μ_{gas}[f]	y_{eff}[g]
1	0.27 ± 0.15	20.5 ± 0.5	< 1.2	< -1.22	< 8.20	2.7 ± 1.7	0.85 ± 0.12	< 0.027
2	0.71 ± 0.17	13.9 ± 0.3	1.4 ± 0.2	$-1.00^{+0.07}_{-0.09}$	$8.33^{+0.07}_{-0.07}$	2.1 ± 0.6	0.63 ± 0.12	0.013 ± 0.003
3	1.5 ± 0.3	18.7 ± 0.4	2.7 ± 0.3	$-0.85^{+0.05}_{-0.05}$	$8.42^{+0.06}_{-0.05}$	3.2 ± 1.1	0.48 ± 0.19	0.010 ± 0.002
4	2.6 ± 0.4	15.9 ± 0.4	2.6 ± 0.3	$-0.78^{+0.05}_{-0.05}$	$8.46^{+0.06}_{-0.05}$	4.0 ± 0.9	0.33 ± 0.12	0.007 ± 0.001
5	4.1 ± 0.6	24.3 ± 0.5	5.3 ± 0.4	$-0.66^{+0.03}_{-0.04}$	$8.52^{+0.06}_{-0.05}$	6.6 ± 1.1	0.36 ± 0.10	0.009 ± 0.002
6	10.5 ± 5.4	27.0 ± 0.4	7.4 ± 0.3	$-0.56^{+0.02}_{-0.02}$	$8.58^{+0.06}_{-0.04}$	13.1 ± 5.6	0.22 ± 0.11	0.007 ± 0.001

[a]Mean and standard deviation of stellar mass from SED fitting; we use a Chabrier (2003) IMF.

[b]Fluxes of Hα and [N II] from the composite spectra.

[c] $N2 \equiv \log(F_{[\mathrm{NII}]}/F_{H\alpha})$.

[d]Oxygen abundance from $N2$, using the calibration of Pettini & Pagel (2004).

[e]Mean and standard devation of the baryonic mass $M_{\mathrm{gas}} + M_\star$, with gas masses determined from the Schmidt law as described in the text.

[f]Mean and standard deviation of the gas fraction $\mu = M_{\mathrm{gas}}/(M_{\mathrm{gas}} + M_\star)$.

[g]Effective yield $y_{\mathrm{eff}} = Z/\ln(1/\mu)$.

Chapter 6

Epilogue

Several unresolved questions remain following this thesis work. We briefly discuss two of them here, because new or upcoming advances in instrumentation should allow them to be addressed in ways that were not possible with the current work.

The first involves the uncertain interpretation of the spatially resolved and tilted emission lines seen in our sample. As we have discussed, such lines could be produced either by large rotating disks or by galactic mergers; the limited spatial resolution of ground-based spectra makes it difficult or impossible to differentiate between the two possibilities. As the two scenarios have very different implications for the maturity of the galaxies involved, the question is an important one. Kinematic data at high angular resolution, preferably covering an entire galaxy, are required in order to construct realistic maps of the velocity field. We anticipate that such data will be obtainable for at least the brighter galaxies in our sample with the new near-IR integral field spectrographs with adaptive optics, which are now coming into use on large telescopes. Our pilot observations are scheduled for October 2005 with OSIRIS, the near-IR integral field spectrograph on the Keck telescope; we will use this time to follow up on some of the more promising candidates in the current sample.

Another unresolved question involves the ratios of the nebular emission lines in the $z \sim 2$ galaxies, and the absolute calibration of the metallicity scale. As was shown in Chapter 4, the ratios of the nebular emission lines in the $z \sim 2$ galaxies appear to be significantly different from those found in both local normal galaxies and local starbursts. Some combination of an increase in electron density and a harder ionzing spectrum probably explains the differences, but with our current data it is not possible to confirm this. Further work on these issues will require a larger sample of galaxies with the full set of emission lines; such data will come from spectrographs that can observe the full set of lines simultaneously (such as GNIRS, on the Gemini telescope), or from the multiobject near-IR spectrographs in development, which will allow us to observe many objects in multiple bands in a reasonable amount of time. In combination with improved photoionization modeling, such observations will undoubtedly lead to an improved understanding of both the physical conditions in the H II regions and of the abundances of the $z \sim 2$ galaxies.

Bibliography

Adelberger, K. L. 2002, PhD thesis, California Institute of Technology

Adelberger, K. L., Shapley, A. E., Steidel, C. C., Pettini, M., Erb, D. K., & Reddy, N. A. 2005a, ApJ, in press

Adelberger, K. L. & Steidel, C. C. 2000, ApJ, 544, 218

Adelberger, K. L., Steidel, C. C., Giavalisco, M., Dickinson, M., Pettini, M., & Kellogg, M. 1998, ApJ, 505, 18

Adelberger, K. L., Steidel, C. C., Pettini, M., Shapley, A. E., Reddy, N. A., & Erb, D. K. 2005b, ApJ, 619, 697

Adelberger, K. L., Steidel, C. C., Shapley, A. E., Hunt, M. P., Erb, D. K., Reddy, N. A., & Pettini, M. 2004, ApJ, 607, 226

Adelberger, K. L., Steidel, C. C., Shapley, A. E., & Pettini, M. 2003, ApJ, 584, 45

Asplund, M., Grevesse, N., Sauval, A. J., Allende Prieto, C., & Kiselman, D. 2004, A&A, 417, 751

Böhm, A., Ziegler, B. L., Saglia, R. P., Bender, R., Fricke, K. J., Gabasch, A., Heidt, J., Mehlert, D., Noll, S., & Seitz, S. 2004, A&A, 420, 97

Baker, A. J., Tacconi, L. J., Genzel, R., Lehnert, M. D., & Lutz, D. 2004, ApJ, 604, 125

Baldwin, J. A., Phillips, M. M., & Terlevich, R. 1981, PASP, 93, 5

Barmby, P., Huang, J.-S., Fazio, G. G., Surace, J. A., Arendt, R. G., Hora, J. L., Pahre, M. A., Adelberger, K. L., Eisenhardt, P., Erb, D. K., Pettini, M., Reach, W. T., Reddy, N. A., Shapley, A. E., Steidel, C. C., Stern, D., Wang, Z., & Willner, S. P. 2004, ApJS, 154, 97

Baugh, C. M., Cole, S., Frenk, C. S., & Lacey, C. G. 1998, ApJ, 498, 504

Begelman, M. C. & Nath, B. B. 2005, MNRAS, 625

Bell, E. F. & de Jong, R. S. 2000, MNRAS, 312, 497

Bell, E. F. & Kennicutt, R. C. 2001, ApJ, 548, 681

Bell, E. F., McIntosh, D. H., Katz, N., & Weinberg, M. D. 2003, ApJS, 149, 289

Benson, A. J., Bower, R. G., Frenk, C. S., Lacey, C. G., Baugh, C. M., & Cole, S. 2003, ApJ, 599, 38

Bernardi, M., Sheth, R. K., Annis, J., Burles, S., Eisenstein, D. J., Finkbeiner, D. P., Hogg, D. W., Lupton, R. H., Schlegel, D. J., SubbaRao, M., Bahcall, N. A., Blakeslee, J. P., Brinkmann, J., Castander, F. J., Connolly, A. J., Csabai, I., Doi, M., Fukugita, M., Frieman, J., Heckman, T., Hennessy, G. S., Ivezić, Ž., Knapp, G. R., Lamb, D. Q., McKay, T., Munn, J. A., Nichol, R., Okamura, S., Schneider, D. P., Thakar, A. R., & York, D. G. 2003, AJ, 125, 1849

Blain, A. W., Smail, I., Ivison, R. J., & Kneib, J.-P. 1999, MNRAS, 302, 632

Bresolin, F., Schaerer, D., González Delgado, R. M., & Stasińska, G. 2005, A&A, in press, astro-ph/0506088

Brinchmann, J., Charlot, S., White, S. D. M., Tremonti, C., Kauffmann, G., Heckman, T., & Brinkmann, J. 2004, MNRAS, 351, 1151

Brinchmann, J. & Ellis, R. S. 2000, ApJ, 536, L77

Bruzual, G. & Charlot, S. 2003, MNRAS, 344, 1000

Buat, V., Boselli, A., Gavazzi, G., & Bonfanti, C. 2002, A&A, 383, 801

Calzetti, D. 1997, AJ, 113, 162

Calzetti, D., Armus, L., Bohlin, R. C., Kinney, A. L., Koornneef, J., & Storchi-Bergmann, T. 2000, ApJ, 533, 682

Calzetti, D., Harris, J., Gallagher, J. S., Smith, D. A., Conselice, C. J., Homeier, N., & Kewley, L. 2004, AJ, 127, 1405

Chabrier, G. 2003, PASP, 115, 763

Chapman, S. C., Blain, A. W., Ivison, R. J., & Smail, I. 2003, Nature, 422, 695

Chapman, S. C., Blain, A. W., Smail, I., & Ivison, R. J. 2005, ApJ, 622, 772

Chapman, S. C., Scott, D., Steidel, C. C., Borys, C., Halpern, M., Morris, S. L., Adelberger, K. L., Dickinson, M., Giavalisco, M., & Pettini, M. 2000, MNRAS, 319, 318

Charlot, S. & Longhetti, M. 2001, MNRAS, 323, 887

Cimatti, A., Daddi, E., Mignoli, M., Pozzetti, L., Renzini, A., Zamorani, G., Broadhurst, T., Fontana, A., Saracco, P., Poli, F., Cristiani, S., D'Odorico, S., Giallongo, E., Gilmozzi, R., & Menci, N. 2002, A&A, 381, L68

Colina, L., Arribas, S., & Monreal-Ibero, A. 2005, ApJ, 621, 725

Colina, L., Bohlin, R. C., & Castelli, F. 1996, Absolute Flux Calibration Spectrum of Vega, (STScI Rep. OSG-CAL-96-01; Baltimore: STScI)

Conselice, C. J., Gallagher, J. S., Calzetti, D., Homeier, N., & Kinney, A. 2000, AJ, 119, 79

Conselice, C. J., Grogin, N. A., Jogee, S., Lucas, R. A., Dahlen, T., de Mello, D., Gardner, J. P., Mobasher, B., & Ravindranath, S. 2004, ApJ, 600, L139

Daddi, E., Cimatti, A., Renzini, A., Fontana, A., Mignoli, M., Pozzetti, L., Tozzi, P., & Zamorani, G. 2004, ApJ, 617, 746

de Mello, D. F., Schaerer, D., Heldmann, J., & Leitherer, C. 1998, ApJ, 507, 199

Dekel, A. & Woo, J. 2003, MNRAS, 344, 1131

Denicoló, G., Terlevich, R., & Terlevich, E. 2002, MNRAS, 330, 69

Di Matteo, T., Croft, R. A. C., Springel, V., & Hernquist, L. 2003, ApJ, 593, 56

Di Matteo, T., Springel, V., & Hernquist, L. 2005, Nature, 433, 604

Dickinson, M. 2000, Philosophical Transactions of the Royal Society of London, Series A, 358, 2001

Dickinson, M., Papovich, C., Ferguson, H. C., & Budavári, T. 2003, ApJ, 587, 25

Dopita, M. A. 2005, in AIP Conf. Proc. 761: The Spectral Energy Distributions of Gas-Rich Galaxies: Confronting Models with Data, 203, astro-ph/0502339

Ellison, S. L., Songaila, A., Schaye, J., & Pettini, M. 2000, AJ, 120, 1175

Erb, D. K., Shapley, A. E., Pettini, M., Steidel, C. C., Reddy, N. A., & Adelberger, K. L. 2005a, in preparation

Erb, D. K., Shapley, A. E., Steidel, C. C., Pettini, M., Adelberger, K. L., Hunt, M. P., Moorwood, A. F. M., & Cuby, J. 2003, ApJ, 591, 101

Erb, D. K., Steidel, C. C., Shapley, A. E., Pettini, M., & Adelberger, K. L. 2004, ApJ, 612, 122

Erb, D. K., Steidel, C. C., Shapley, A. E., Pettini, M., Reddy, N. A., & Adelberger, K. L. 2005b, in preparation

Förster Schreiber, N. M., van Dokkum, P. G., Franx, M., Labbé, I., Rudnick, G., Daddi, E., Illingworth, G. D., Kriek, M., Moorwood, A. F. M., Rix, H.-W., Röttgering, H., Trujillo, I., van der Werf, P., van Starkenburg, L., & Wuyts, S. 2004, ApJ, 616, 40

Faber, S. M. & Jackson, R. E. 1976, ApJ, 204, 668

Fan, X. et al. 2001, AJ, 121, 54

Feigelson, E. D. & Nelson, P. I. 1985, ApJ, 293, 192

Ferrarese, L. & Merritt, D. 2000, ApJ, 539, L9

Franx, M., Labbé, I., Rudnick, G., van Dokkum, P. G., Daddi, E., Förster Schreiber, N. M., Moorwood, A., Rix, H., Röttgering, H., van de Wel, A., van der Werf, P., & van Starkenburg, L. 2003, ApJ, 587, L79

Fruchter, A. S. & Hook, R. N. 2002, PASP, 114, 144

Garnett, D. R. 2002, ApJ, 581, 1019

Garnett, D. R., Kennicutt, R. C., & Bresolin, F. 2004, ApJ, 607, L21

Garnett, D. R., Shields, G. A., Skillman, E. D., Sagan, S. P., & Dufour, R. J. 1997, ApJ, 489, 63

Genzel, R., Baker, A. J., Tacconi, L. J., Lutz, D., Cox, P., Guilloteau, S., & Omont, A. 2003, ApJ, 584, 633

Giavalisco, M. & Dickinson, M. 2001, ApJ, 550, 177

Giavalisco, M., Steidel, C. C., & Macchetto, F. D. 1996, ApJ, 470, 189

Giavalisco, M. et al. 2004, ApJ, 600, L93

Glazebrook, K., Blake, C., Economou, F., Lilly, S., & Colless, M. 1999, MNRAS, 306, 843

Haehnelt, M. G., Steinmetz, M., & Rauch, M. 1998, ApJ, 495, 647

Heckman, T. M. 2002, in ASP Conf. Ser. 254: Extragalactic Gas at Low Redshift, 292

Heckman, T. M., Armus, L., & Miley, G. K. 1990, ApJS, 74, 833

Hernquist, L. & Springel, V. 2003, MNRAS, 341, 1253

Hopkins, A. M., Connolly, A. J., & Szalay, A. S. 2000, AJ, 120, 2843

Hopkins, A. M., Miller, C. J., Nichol, R. C., Connolly, A. J., Bernardi, M., Gómez, P. L., Goto, T., Tremonti, C. A., Brinkmann, J., Ivezić, Ž., & Lamb, D. Q. 2003, ApJ, 599, 971

Johnson, K. E., Leitherer, C., Vacca, W. D., & Conti, P. S. 2000, AJ, 120, 1273

Juneau, S., Glazebrook, K., Crampton, D., McCarthy, P. J., Savaglio, S., Abraham, R., Carlberg, R. G., Chen, H., Le Borgne, D., Marzke, R. O., Roth, K., Jørgensen, I., Hook, I., & Murowinski, R. 2005, ApJ, 619, L135

Kauffmann, G., Heckman, T. M., Tremonti, C., Brinchmann, J., Charlot, S., White, S. D. M., Ridgway, S. E., Brinkmann, J., Fukugita, M., Hall, P. B., Ivezić, Ž., Richards, G. T., & Schneider, D. P. 2003, MNRAS, 346, 1055

Kennicutt, R. C. 1998a, ARAA, 36, 189

—. 1998b, ApJ, 498, 541

Kennicutt, R. C., Bresolin, F., & Garnett, D. R. 2003, ApJ, 591, 801

Kewley, L. J. & Dopita, M. A. 2002, ApJS, 142, 35

Kewley, L. J., Dopita, M. A., Sutherland, R. S., Heisler, C. A., & Trevena, J. 2001a, ApJ, 556, 121

Kewley, L. J. & Ellison, S. J. 2005, in preparation

Kewley, L. J., Heisler, C. A., Dopita, M. A., & Lumsden, S. 2001b, ApJS, 132, 37

Kobulnicky, H. A. & Gebhardt, K. 2000, AJ, 119, 1608

Kobulnicky, H. A., Kennicutt, R. C., & Pizagno, J. L. 1999, ApJ, 514, 544

Kobulnicky, H. A. & Kewley, L. J. 2004, ApJ, 617, 240

Kobulnicky, H. A. & Koo, D. C. 2000, ApJ, 545, 712

Kobulnicky, H. A., Willmer, C. N. A., Phillips, A. C., Koo, D. C., Faber, S. M., Weiner, B. J., Sarajedini, V. L., Simard, L., & Vogt, N. P. 2003, ApJ, 599, 1006

Koekemoer, A. M. et al. 2002, "HST Dither Handbook," Version 2.0, (Baltimore: STScI)

Kroupa, P. 2001, MNRAS, 322, 231

Lamareille, F., Mouhcine, M., Contini, T., Lewis, I., & Maddox, S. 2004, MNRAS, 350, 396

Lançon, A., Goldader, J. D., Leitherer, C., & Delgado, R. M. G. . 2001, ApJ, 552, 150

Larson, R. B. 1974, MNRAS, 169, 229

Lavalley, M., Isobe, T., & Feigelson, E. 1992, in ASP Conf. Ser. 25: Astronomical Data Analysis Software and Systems I, Vol. 1, 245

Law, D. L., Steidel, C. C., & Erb, D. K. 2005, AJ, submitted

Lehnert, M. D. & Heckman, T. M. 1996, ApJ, 472, 546

Lehnert, M. D., Heckman, T. M., & Weaver, K. A. 1999, ApJ, 523, 575

Leitherer, C., Leão, J. R. S., Heckman, T. M., Lennon, D. J., Pettini, M., & Robert, C. 2001, ApJ, 550,

724

Leitherer, C., Schaerer, D., Goldader, J. D., Delgado, R. M. G., Robert, C., Kune, D. F., de Mello, D. F., Devost, D., & Heckman, T. M. 1999, ApJS, 123, 3

Leitherer, C., Vacca, W. D., Conti, P. S., Filippenko, A. V., Robert, C., & Sargent, W. L. W. 1996, ApJ, 465, 717

Lemoine-Busserolle, M., Contini, T., Pelló, R., Le Borgne, J.-F., Kneib, J.-P., & Lidman, C. 2003, A&A, 397, 839

Lequeux, J., Peimbert, M., Rayo, J. F., Serrano, A., & Torres-Peimbert, S. 1979, A&A, 80, 155

Liang, Y. C., Hammer, F., Flores, H., Elbaz, D., Marcillac, D., & Cesarsky, C. J. 2004, A&A, 423, 867

Lilly, S. J., Carollo, C. M., & Stockton, A. N. 2003, ApJ, 597, 730

Madau, P., Pozzetti, L., & Dickinson, M. 1998, ApJ, 498, 106

Magorrian, J., Tremaine, S., Richstone, D., Bender, R., Bower, G., Dressler, A., Faber, S. M., Gebhardt, K., Green, R., Grillmair, C., Kormendy, J., & Lauer, T. 1998, AJ, 115, 2285

Maier, C., Meisenheimer, K., & Hippelein, H. 2004, A&A, 418, 475

Marconi, A. & Hunt, L. K. 2003, ApJ, 589, L21

Martin, C. L. 1999, ApJ, 513, 156

Martin, C. L. 2003, in Revista Mexicana de Astronomia y Astrofisica Conference Series, 56–59

—. 2005, ApJ, 621, 227

Martin, C. L. & Kennicutt, R. C. 2001, ApJ, 555, 301

Martin, C. L., Kobulnicky, H. A., & Heckman, T. M. 2002, ApJ, 574, 663

Matthews, K. & Soifer, B. T. 1994, in ASSL Vol. 190: Astronomy with Arrays, The Next Generation, 239

McGaugh, S. S. & de Blok, W. J. G. 1997, ApJ, 481, 689

McGaugh, S. S., Schombert, J. M., Bothun, G. D., & de Blok, W. J. G. 2000, ApJ, 533, L99

McLean, I. S., Becklin, E. E., Bendiksen, O., Brims, G., Canfield, J., Figer, D. F., Graham, J. R., Hare, J., Lacayanga, F., Larkin, J. E., Larson, S. B., Levenson, N., Magnone, N., Teplitz, H., & Wong, W. 1998, in Proc. SPIE, Infrared Astronomical Instrumentation, Albert M. Fowler; Ed., Vol. 3354, 566–578

Metevier, A. J., Koo, D. C., Simard, L., & Phillips, A. C. 2004, ApJ, submitted

Minniti, D. 1996, ApJ, 459, 175

Mo, H. J. & Mao, S. 2004, MNRAS, 353, 829

Mo, H. J., Mao, S., & White, S. D. M. 1998, MNRAS, 295, 319

—. 1999, MNRAS, 304, 175

Moorwood, A., Cuby, J.-G., Biereichel, P., Brynnel, J., Delabre, B., Devillard, N., van Dijsseldonk, A., Finger, G., Gemperlein, H., Gilmozzi, R., Herlin, T., Huster, G., Knudstrup, J., Lidman, C., Lizon, J.-L., Mehrgan, H., Meyer, M., Nicolini, G., Petr, M., Spyromilio, J., & Stegmeier, J. 1998, The Messenger, 94, 7

Moorwood, A. F. M., van der Werf, P. P., Cuby, J. G., & Oliva, E. 2003, Proceedings of the Workshop on The Mass of Galaxies at Low and High Redshift, eds. R. Bender and A. Renzini, Springer-Verlag, p. 302

Murray, N., Quataert, E., & Thompson, T. A. 2005, ApJ, 618, 569

Naab, T. & Ostriker, J. P. 2005, MNRAS, submitted, astro-ph/0505594

Nagamine, K., Springel, V., Hernquist, L., & Machacek, M. 2004, MNRAS, 350, 385

Nandra, K., Mushotzky, R. F., Arnaud, K., Steidel, C. C., Adelberger, K. L., Gardner, J. P., Teplitz, H. I., & Windhorst, R. A. 2002, ApJ, 576, 625

Oke, J. B., Cohen, J. G., Carr, M., Cromer, J., Dingizian, A., Harris, F. H., Labrecque, S., Lucinio, R., Schaal, W., Epps, H., & Miller, J. 1995, PASP, 107, 375

Osterbrock, D. E. 1989, Astrophysics of Gaseous Nebulae and Active Galactic Nuclei, (Mill Valley: University Science Books)

Pagel, B. E. J. & Patchett, B. E. 1975, MNRAS, 172, 13

Papovich, C., Dickinson, M., & Ferguson, H. C. 2001, ApJ, 559, 620

Pei, Y. C. 1995, ApJ, 438, 623

Pettini, M. & Pagel, B. E. J. 2004, MNRAS, 348, L59

Pettini, M., Rix, S. A., Steidel, C. C., Adelberger, K. L., Hunt, M. P., & Shapley, A. E. 2002, ApJ, 569, 742

Pettini, M., Shapley, A. E., Steidel, C. C., Cuby, J., Dickinson, M., Moorwood, A. F. M., Adelberger, K. L., & Giavalisco, M. 2001, ApJ, 554, 981

Pettini, M., Steidel, C. C., Adelberger, K. L., Dickinson, M., & Giavalisco, M. 2000, ApJ, 528, 96

Pizzella, A., Corsini, E. M., Dalla Bontà, E., Sarzi, M., Coccato, L., & Bertola, F. 2005, ApJ, in press, astro-ph/0503649

Raimann, D., Storchi-Bergmann, T., Bica, E., Melnick, J., & Schmitt, H. 2000, MNRAS, 316, 559

Reddy, N. A., Erb, D. K., Steidel, C. C., Shapley, A. E., Adelberger, K. L., & Pettini, M. 2005, ApJ, in press, astro-ph/0507264

Reddy, N. A. & Steidel, C. C. 2004, ApJ, 603, L13

Renzini, A. 2005, in IMF@50: The Initial Mass Function, 50 Years Later, in press, astro-ph/0410295

Rix, H., Guhathakurta, P., Colless, M., & Ing, K. 1997, MNRAS, 285, 779

Rix, S. A., Pettini, M., Leitherer, C., Bresolin, F., Kudritzki, R., & Steidel, C. C. 2004, ApJ, 615, 98

Robertson, B., Hernquist, L., Cox, T. J., Di Matteo, T., Hopkins, P. F., Martini, P., & Springel, V. 2005, ApJ, submitted, astro-ph/0506038

Rudnick, G., Rix, H., Franx, M., Labbé, I., Blanton, M., Daddi, E., Förster Schreiber, N. M., Moorwood, A., Röttgering, H., Trujillo, I., van de Wel, A., van der Werf, P., van Dokkum, P. G., & van Starkenburg, L. 2003, ApJ, 599, 847

Rupke, D. S., Veilleux, S., & Sanders, D. B. 2005, ApJS, in press, astro-ph/0506611

Salpeter, E. E. 1955, ApJ, 121, 161

Salzer, J. J., Lee, J. C., Melbourne, J., Hinz, J. L., Alonso-Herrero, A., & Jangren, A. 2005, ApJ, 624, 661

Schmidt, M. 1959, ApJ, 129, 243

—. 1963, ApJ, 137, 758

Schmidt, M., Schneider, D. P., & Gunn, J. E. 1995, AJ, 110, 68

Searle, L. & Sargent, W. L. W. 1972, ApJ, 173, 25

Shapley, A. E., Coil, A. L., Ma, C.-P., & Bundy, K. 2005a, ApJ, submitted

Shapley, A. E., Erb, D. K., Pettini, M., Steidel, C. C., & Adelberger, K. L. 2004, ApJ, 612, 108

Shapley, A. E., Steidel, C. C., Adelberger, K. L., Dickinson, M., Giavalisco, M., & Pettini, M. 2001, ApJ, 562, 95

Shapley, A. E., Steidel, C. C., Erb, D. K., Reddy, N. A., Adelberger, K. L., Pettini, M., Barmby, P., & Huang, J. 2005b, ApJ, 626, 698

Shapley, A. E., Steidel, C. C., Pettini, M., & Adelberger, K. L. 2003, ApJ, 588, 65

Simard, L. & Pritchet, C. J. 1998, ApJ, 505, 96

Simcoe, R. A., Sargent, W. L. W., & Rauch, M. 2004, ApJ, 606, 92

Skillman, E. D., Kennicutt, R. C., & Hodge, P. W. 1989, ApJ, 347, 875

Smail, I., Chapman, S. C., Ivison, R. J., Blain, A. W., Takata, T., Heckman, T. M., Dunlop, J. S., & Sekiguchi, K. 2003, MNRAS, 342, 1185

Somerville, R. S. & Primack, J. R. 1999, MNRAS, 310, 1087

Spergel, D. N., Verde, L., Peiris, H. V., Komatsu, E., Nolta, M. R., Bennett, C. L., Halpern, M., Hinshaw, G., Jarosik, N., Kogut, A., Limon, M., Meyer, S. S., Page, L., Tucker, G. S., Weiland, J. L., Wollack, E., & Wright, E. L. 2003, ApJS, 148, 175

Stasińska, G. 2005, A&A, 434, 507

Steidel, C. C., Adelberger, K. L., Giavalisco, M., Dickinson, M., & Pettini, M. 1999, ApJ, 519, 1

Steidel, C. C., Adelberger, K. L., Shapley, A. E., Erb, D. K., Reddy, N. A., & Pettini, M. 2005, ApJ, 626, 44

Steidel, C. C., Adelberger, K. L., Shapley, A. E., Pettini, M., Dickinson, M., & Giavalisco, M. 2003, ApJ, 592, 728

Steidel, C. C., Giavalisco, M., Pettini, M., Dickinson, M., & Adelberger, K. L. 1996, ApJ, 462, L17

Steidel, C. C. & Hamilton, D. 1993, AJ, 105, 2017

Steidel, C. C., Pettini, M., & Hamilton, D. 1995, AJ, 110, 2519

Steidel, C. C., Shapley, A. E., Pettini, M., Adelberger, K. L., Erb, D. K., Reddy, N. A., & Hunt, M. P. 2004, ApJ, 604, 534

Storchi-Bergmann, T., Calzetti, D., & Kinney, A. L. 1994, ApJ, 429, 572

Strickland, D. K., Heckman, T. M., Colbert, E. J. M., Hoopes, C. G., & Weaver, K. A. 2004, ApJ, 606, 829

Sullivan, M., Treyer, M. A., Ellis, R. S., Bridges, T. J., Milliard, B., & Donas, J. 2000, MNRAS, 312, 442

Swaters, R. A., Madore, B. F., van den Bosch, F. C., & Balcells, M. 2003, ApJ, 583, 732

Swinbank, A. M., Smail, I., Chapman, S. C., Blain, A. W., Ivison, R. J., & Keel, W. C. 2004, ApJ, 617, 64

Tantalo, R., Chiosi, C., & Bressan, A. 1998, A&A, 333, 419

Teplitz, H. I., Malkan, M., & McLean, I. S. 1998, ApJ, 506, 519

Tiede, G. P. & Terndrup, D. M. 1999, AJ, 118, 895

Tremaine, S., Gebhardt, K., Bender, R., Bower, G., Dressler, A., Faber, S. M., Filippenko, A. V., Green, R., Grillmair, C., Ho, L. C., Kormendy, J., Lauer, T. R., Magorrian, J., Pinkney, J., & Richstone, D. 2002, ApJ, 574, 740

Tremonti, C. A., Heckman, T. M., Kauffmann, G., Brinchmann, J., Charlot, S., White, S. D. M., Seibert, M., Peng, E. W., Schlegel, D. J., Uomoto, A., Fukugita, M., & Brinkmann, J. 2004, ApJ, 613, 898

Tresse, L., Maddox, S. J., Le Fèvre, O., & Cuby, J.-G. 2002, MNRAS, 337, 369

van den Bergh, S. 1962, AJ, 67, 486

—. 2001, AJ, 122, 621

van Dokkum, P. G., Franx, M., Förster Schreiber, N. M., Illingworth, G. D., Daddi, E., Knudsen, K. K., Labbé, I., Moorwood, A., Rix, H., Röttgering, H., Rudnick, G., Trujillo, I., van der Werf, P., van der Wel, A., van Starkenburg, L., & Wuyts, S. 2004, ApJ, 611, 703

van Dokkum, P. G., Kriek, M., Rodgers, B., Franx, M., & Puxley, P. 2005, ApJ, 622, L13

Veilleux, S. & Osterbrock, D. E. 1987, ApJS, 63, 295

Vogt, N. P., Forbes, D. A., Phillips, A. C., Gronwall, C., Faber, S. M., Illingworth, G. D., & Koo, D. C. 1996, ApJ, 465, L15

Vogt, N. P., Phillips, A. C., Faber, S. M., Gallego, J., Gronwall, C., Guzman, R., Illingworth, G. D., Koo, D. C., & Lowenthal, J. D. 1997, ApJ, 479, L121

Weatherley, S. J. & Warren, S. J. 2003, MNRAS, 345, L29

Wilkinson, M. I. & Evans, N. W. 1999, MNRAS, 310, 645

Wilson, J. C., Eikenberry, S. S., Henderson, C. P., Hayward, T. L., Carson, J. C., Pirger, B., Barry, D. J., Brandl, B. R., Houck, J. R., Fitzgerald, G. J., & Stolberg, T. M. 2003, in Instrument Design and Performance for Optical/Infrared Ground-based Telescopes. Edited by Iye, Masanori; Moorwood, Alan F. M. Proceedings of the SPIE, Volume 4841, 451–458

Worthey, G. 1998, PASP, 110, 888

Yan, L., McCarthy, P. J., Freudling, W., Teplitz, H. I., Malumuth, E. M., Weymann, R. J., & Malkan, M. A. 1999, ApJ, 519, L47

Zaritsky, D. 1999, in ASP Conf. Ser. 165: The Third Stromlo Symposium: The Galactic Halo, 34

Zaritsky, D., Kennicutt, R. C., & Huchra, J. P. 1994, ApJ, 420, 87

www.ingramcontent.com/pod-product-compliance
Lightning Source LLC
Chambersburg PA
CBHW081123170526
45165CB00008B/2529